The Call of the Cranes

Bernhard Wessling

The Call of the Cranes

Expeditions into a Mysterious World

 Springer

Bernhard Wessling
BWITB
Jersbek, Germany

ISBN 978-3-030-98282-9   ISBN 978-3-030-98283-6   (eBook)
https://doi.org/10.1007/978-3-030-98283-6

*For George, without whom this book wouldn't even be half as interesting;
for my grandchildren, whose inquiries prompted me to write it;
for my partner, she supports me so much.*

Foreword

When a pair of Sandhill Cranes that nested in a small wetland near our home lost their 3-week-old juvenile to predators in the middle of the night, the parents called for the remainder of that night, and frequently throughout the next few days. They seemed to grieve. Those that work with captive cranes know the great variation in their personalities. The more one experiences cranes in the wild, the more fascinating they become. After all, these tallest of birds that fly have graced earth's landscapes for millions of years before humans appeared. Evolution has had plenty of time to perfect a remarkable product.

Dr. Bernhard Wessling, a nature lover and an accomplished chemist, became interested in cranes in 1981, when a pair of Eurasian Cranes established themselves in a nature preserve near Hamburg, Germany. He observed them frequently and recorded extensive notes about their behavior. In the following years, more pairs established breeding territories in the wetlands and meadows. Bernhard wanted to know more about them. To protect the cranes, a rule was created that humans were not allowed to leave footpaths. Because it was impossible to capture and color band the cranes, Bernhard recorded their calls and, through analysis of duets performed by pair members, he was able to identify every bird. Year after year, by recording the duets, he was able to peer into the lives of both individuals and pairs. This book reveals his remarkable discoveries, collected in the wild, about the behavior of these complicated creatures (Fig. 1).

At a conference about cranes in Germany in 1996, I was enthralled by Bernhard's findings, and suggested he use his techniques to learn more about the rarest of cranes, the Whooping Cranes in North America, and the Red-Crowned Cranes in the Orient—both close relatives of the abundant Eurasian Cranes. Whooping Cranes numbered in the low hundreds. Vocal

Fig. 1 Whooping and Sandhill Cranes resting at Platte River, Nebraska, during their migration to the South: It is very rare to see so many Whooping Cranes resting together during migration. Photo taken on Nov 6, 2021 (© Crane Trust by Colleen Childers)

identification avoided the risks associated with capturing cranes to band them. Dr. Wessling's recordings of wild Whooping Crane vocalizations have been used during the raising and flight training of captive-reared Whooping Cranes to communicate with them and have them follow ultralight aircrafts to learn their migration routes across the USA. Thus, he has made a significant contribution to the reintroduction of a migratory flock of Whooping Cranes east of the Mississippi River.

And I had a hunch that the non-migratory Red-Crowned Cranes in Japan were perhaps genetically distinct from their migratory cousins on mainland Asia. Perhaps their call might provide more evidence to answer an important question for conservation. Bernhard's adventures and findings in Texas, Japan, and South Korea (in the DMZ at the border to North Korea) are outstanding.

Beyond the many obvious questions that individual recognition answered, Bernhard's analytical mind ventures into the questions about the ability of cranes to reason and not simply to respond. That is where this book becomes unique and thought-provoking and brought me back to thinking about the wild Sandhills near my home.

One summer, I fed them corn every morning some distance from our house along the driveway that ran through the marsh. The cranes were standoffish and never came near us. One morning, I forgot to feed them. They walked to the house, climbed up the steps of the porch, stood beside the kitchen door, and called. As well as getting the message, I also realized that cranes not only respond, but they can also think!

Enjoy the read. And thank you, Bernhard, for your substantial contribution to our understanding of these remarkable birds that migrate across continents, dance, duet, lavish care on their young, and still survive, despite the ways in which we modern humans have impacted their ancient lives.

International Crane Foundation   George Archibald
Baraboo, WI, USA

Foreword

Sy Montgomery, naturalist and author of 31 books for both adults and children (including *The Soul of an Octopus: A Surprising Exploration into the Wonder of Consciousness*, which was featured on the *New York Times* bestseller list).

The Call of the Cranes is a mesmerizing, vivid, lyrical, and revelatory book. Full of beauty, suspense, and insight, it is not just about a beautiful and mysterious bird—though this alone makes these pages thrilling reading. But Bernhard Wessling brings us even more. He has spent many years conducting studies on the intelligence and behavior of four species of cranes in the wild, including a new method he developed to identify crane individuals and pairs in his study areas without disturbing them at all: from a distance, by analyzing their voice. This book is a testament to the joy and dedication that ignites when we deeply connect with individuals of other species, when we enter and inhabit their world. And it is also a call to arms, inspiring us to summon the courage we need to save the cranes—and all the other species threatened by us humans.

Syracuse University   Sy Montgomery
Hancock, NH, USA

Preface

It was a long way from the densely populated, ugly, and heavily polluted Ruhr region, where I grew up and studied, to the Duvenstedt Brook near Hamburg, where I saw cranes for the first time in my life. Even longer and more arduous was my expedition into the hidden, mysterious world of the cranes, their life, and their way of thinking.

I came into contact with the issues of environmental pollution and threats to nature at a very early age. As a child, I often noticed, when our family of eight's laundry was hanging outside in the garden, how a cloud of soot would rise from the chimneys of the nearby coking plant in Herne, settling in our backyard and leaving ugly black stains on the clean clothes. As a teenager, I loved the late autumn evenings, when the dense fog forced the then still small number of cars to proceed at a walking pace, while I drove my bike, to which I had attached powerful lamps so as to conjure up mighty cones of light within the fog, which was, in fact, smog.

In 1971, as a third-semester chemistry student, I responded to a blackboard notice seeking chemists to analyze illegally dumped barrels.[1] Most of these contained cyanide compounds, and, to a lesser extent, other substances, and some contained sulfuric acid. The barrels had been dumped into a specially dug hole, which had gradually filled with water. The sulfuric acid barrels rotted first, so that this "pond" was, by now, strongly acidic, a condition that, with the additional presence of the cyanide salts in the also slowly corroding barrels, had led to the release of hydrogen cyanide gas. Dead animals lay around the pond and floated on the water. It was a "doomsday" scenario. As a university student with no funds, I desperately needed money to make a

---

[1] cf. http://www.spiegel.de/spiegel/print/d-43144036.html

living. The difficult and dangerous job paid well. So, I found myself wearing full respiratory protective gear in sweltering heat during the semester break. Every day for weeks, often enveloped in highly toxic dust clouds, I spent 6–8 h analyzing rotting barrels to see if they contained cyanides ("to the left, to the big barrel mountain") or other less toxic waste salts ("to the right, to the other toxic waste").

The heat was suffocating. Threatening dust clouds passed over us from all directions. The full protective clothing and gas masks required were actually unbearable. This tempted some workers to work without breathing protection. One day, one of them was sitting in front of me on top of his excavator. He wanted me to examine the barrels he was excavating. When he moved his shovel, he accidentally caught a barrel of powder, the rotten barrel shattered, a dust cloud surrounded me and the excavator, and the excavator operator immediately collapsed dead before my eyes right there in the driver's seat. I frantically called for the paramedics, and the worker was rushed to the mobile emergency clinic installed on the site, injected with an antidote within seconds, which revived him, and he was additionally ventilated. The next day, he was back on the excavator, but now wearing a gas mask and full protective clothing. None of the workers refused to take the necessary protective measures from then on. The weeks-long student job shaped my attitude toward environmental protection and, later, nature conservation. A year later, in 1972, the first report of the Club of Rome, "The Limits of Growth," appeared and was hotly debated among us chemistry students. It became increasingly clear to me: We needed to treat this planet and its ecosystems with much more respect. As a chemist, I wanted to make my contribution to this through research.

By the age of about 14, I had already become intensely involved in natural sciences, including astronomy. When I looked into space through my telescope for which I had painstakingly saved up, I felt not only insatiable curiosity and boundless awe, but also a deep-seated fear of the infinity of the universe. I was then struck by severe depression: We are alone on our earth, floating in hostile space—that is how I felt, and it caused me to feel lonely. The situation was only exacerbated by the fact that I had little support within my own family, and thus I became something of a loner.

One day, as I was once again wandering aimlessly through a small forest in Herne, I found a tiny, bluish shimmering feather. I learned that it was a jay feather and put it in a small box. On further rambles, I collected more and more feathers, even including one from an eagle! I attached them to a white piece of cardboard that I hung on the wall in my basement room; I discovered that I enjoyed studying bird feathers and spending time in nature, and, in this way, I found my way out of my fears and my deep depression. Forests and fields had become places of retreat for me, where I could relax and reflect on myself and the world. Nature—which includes those landscapes that are shaped by humans as well as the wild, rugged, hard-to-reach, and lonely areas—has since then been a regular source of relaxation and relief from professional and personal stress for me. (To determine this effect today has required elaborate research, but at least the latest studies from the USA and Japan confirm my personal experience over the past five-plus decades.)

As a young family man, I brought my children into contact with nature from the beginning. In particular, we watched birds and discovered the cranes for ourselves. Together with my growing sons, I came to understand these birds' vulnerability and how difficult it is to protect or restore their habitat, and that nature and species conservation must always go hand in hand with environmental protection. I decided to join the crane conservation program, which I would soon come to lead for about 5 years.

During my intensive observation of the cranes, I discovered that shockingly little was known about the life and behavior of these impressive birds. With their enigmatic nature, they aroused my scientifically trained curiosity and inspired me to research outside of my real profession.

There is unlikely to be another place in the world where free and wild cranes live and breed in such close proximity to humans as the Duvenstedt Brook and the Hansdorf Brook. Both are located on the northern edge of Hamburg, a city of over a million inhabitants, tens of thousands of whom visit the nature reserve every year to hike, relax, and observe nature. (Unfortunately, a not negligible minority of the visitors disturbed the nature reserve with their picnics, Easter egg hunts, and venturing off the paths to take pictures, activities that are sometimes accompanied by poaching and egg theft. This situation has since improved greatly, due to our persistent work.)

Perhaps nowhere else were crane watchers as intensely connected to "their" cranes as we were. The task of the "crane guards," as we called ourselves and were called by the visitors, was to prevent disturbances. So, we did not actually "guard" the cranes, but rather the visitors, at least those who would consciously or unconsciously become troublemakers.

During the breeding season, there were usually two crane guards in the Brook all day every day for one week. Many of us even spent the night there. We got up at the crack of dawn and did not go to sleep until after the "woodcock dash" (this is the name given to the behavior of woodcocks that "dash" at dusk along the edge of the forest or across the meadows in their territory).

From mid-February to mid-November, the cranes are "with us." Until the late 1990s, there were four to six breeding pairs of cranes and a few "bachelors" hanging around our area each year. In the early 2000s up to around 2016, about a dozen crane pairs each occupied a territory. In 2019, in addition to the dozen territorial pairs and other pairs seeking territories, at times, more than 20 juvenile cranes, some as a large group, were in the Brook. One day in May of that year, I saw 65 cranes in a meadow in the core of the Brook, while in May 2020, there were more than 100 (which was not at all the case in 2021). By the way, the territories, in the narrower sense, are no larger than about half a square kilometer, and in some places, they are easy to look into (although most parts are not observable, to the advantage of the cranes). However, the territorial pairs defend a much larger area against other cranes, so the territories include a core zone with a breeding site and a feeding area, as well as a buffer zone.

So, for years—perhaps uniquely in the world—I was able to observe many cranes under open-air conditions just a few minutes away from my home and my workplace. In the spirit of our protection mission, we observed the animals from afar, from outside the flight distance, so that the observation itself did not have any disturbing effect.

I did not conduct behavioral experiments with cranes, but only observed them. However, this does not mean that I can observe and describe "cranes unaffected by humans." Humans restrict the breeding and feeding areas and the mobility of the animals through hiking paths, roads, or agricultural areas. The latter have adapted their behavior, and so one always also observes the birds' reactions to human influences. The behavior of animals in a cultivated landscape like the Brook is certainly not the same as in the wild, in places such as the largely undisturbed Siberian tundra, the mid-Swedish forest, or the Finnish lake landscape, although, in the meantime, some smaller areas in our nature reserve have been left to their natural development again.

It was precisely this circumstance that made the observations particularly appealing: How do cranes deal with situations that are unfamiliar to them? How do they behave when other animals, but especially humans, disturb their breeding or feeding? Those who, like me, have enjoyed observing nature from a young age, no matter in what field, will sooner or later come across strange events. I noticed that "my" cranes behaved differently than I had expected after brushing up on my knowledge of behavioral science and reading contemporary articles and books on cranes. In contrast to what I read and heard, they did not behave stereotypically, not as one would expect according to an inherited behavioral pattern, but like actual personalities, with their own plans and individual traits.

This did not come as a complete surprise to me. Again and again, I had thought about how "thinking" actually goes on, what the material basis of memory is, and how consciousness arises. In this process, I occasionally wondered whether animals' thought processes are really so very different from our own, and it would seem perfectly normal to me if, one day, it were discovered that animals think in a fundamentally similar way to humans, merely—depending on the species—gradually differing from us and from each other. So, I am always eager to read articles or books that report research results on the thinking, intelligence, and consciousness of animals.

I had not expected that I, as a voluntary conservationist, would ever be in a position to contribute my own systematic observations on this subject. But more and more, my observations had turned into real and systematic scientific research. As a nature scientist by education, as a chemist who started during the PhD laboratory work with deep research, and as the one who continued to even perform fundamental research in combination with applied product development in the mid-size chemical company that I ran as CEO and major shareholder, I entered more and more into behavioral research. Still I did not expect that anyone would ever be interested in what I observed and concluded. But when I presented some particularly remarkable observations from my first years on the occasion of the European Crane Conference 1996 in Stralsund, one of the attendees, George Archibald, was listening, the founder of the International Crane Foundation (ICF), famous among crane experts and conservationists all over the world. He motivated me to deepen my studies and, additionally, to pursue them internationally, with crane species beyond the Common Crane (also called the "Eurasian Crane") native to our country and most of Europe in general, but also to parts of Asia. Thus, over time, I actively participated in numerous international projects, conducted crane research parallel both to my main job as a chemical researcher and to what I did as an entrepreneur for the development of my company, and published the results of my work at conferences and in specialist publications.

Since mid-May 2018, my life partner and I can hear crane calls when we wake up in the morning or sometime during the day. When I sit at my desk in my study under the roof, I look out across the landscape during thoughtful pauses. We now live in the immediate vicinity of the Hansdorf Brook on the outskirts of Hamburg. Time and again, cranes fly by at a distance of only 50 or 150 m. Even more often, I hear them calling. Shortly after we moved into the house on the edge of the Brook, I told my then nine-year-old grandson the story of "Romeo and Juliet," the crane pair that readers of this book will get to know better later. Their last nesting site is only about 300 m from our house as the crow flies. It was at that moment that I began thinking about writing this book. It was about the same time when George visited me in my new house and reminded me of his wish from a long time ago that I should write this book. I was lucky that I had written lots of diary entries and had even drafted many raw text chapters during the years with the cranes, and also lots of systematic notes and tables (just as nature scientists do it)—otherwise I would not have been able to write the book as you can now read it. It appeared in March 2020 in German, on the very day when Germany commenced its first lockdown due to the Corona pandemic. This English edition is slightly revised and partially updated where necessary.

Here, I recreate my years-long expeditions into the enigmatic world of the cranes. I describe experiences and observations that have allowed me to solve some of the mysteries that these beautiful birds have presented to us humans for millennia. These mysteries, in turn, lead us to questions about ourselves and our consciousness: How rationally, how consciously do we humans act, and how different is this from the actions and thoughts of animals, specifically, in this instance, cranes?

After realizing that these birds are different from what has been described in the textbooks so far, it is not a long path to the conclusion that we need to think much more broadly about conservation and that we need to act more holistically—based on a deep respect for nature.

Jersbek, Germany   Bernhard Wessling

Acknowledgments

I am grateful to so many people for their help in the making of this book. First, George Archibald, without whose encouragement I would have continued to watch cranes only in the Brook. He lured me into the much bigger international projects. My sons, who accompanied me on many of my early observations, especially my older son, with whom I spent a number of joint crane-watching weeks. My life partner, who read several versions of the manuscript for the German edition, gave me numerous critical suggestions, and also had endless patience with me during the final phase of constantly editing the manuscript until it was finally approved. (I am especially grateful that she has since come to share my passion for cranes and nature.) And then, she developed even more patience during the subsequent phase, when I worked with the editor at Springer Nature and their lector for this English edition. I thank the publisher Goldmann and the editorial office for seeing my book project as a suitable offering to a readership interested in nature, and for their many valuable comments and suggestions. In addition, more thanks go to Goldmann for allowing Springer Nature to publish the book in English, as well as, to be sure, to Springer Nature and its editors for accepting my proposal to publish this edition, which would have not have been in true English if Marc Beschler had not reviewed the various versions that I sent with ever ongoing changes. I would like to thank the photographers Carsten Linde, Kunikazu Momose, Colleen Childers, Ted Thousand, Tom Lynn, Larry Mattney, and Mike Endres, the crane conservation organizations ICF (International Crane Foundation) and Red-Crowned Crane Conservancy Hokkaido, as well as the chief pilot of Operation Migration, Joe Duff, for providing me with photos that complement my own and also with additional information and suggestions. Also, I am very grateful to Jennifer Ackerman and Sy Montgomery: for your inspiring communication and your support for my book. Even more thanks have to go to you, George, for your Foreword and your various detailed suggestions after you read a preliminary version of this English edition, as well as for the initiative you showed in sending me here and there and your motivating support! (Fig. 2)

Fig. 2   I am waiting for the cranes to call (© Bernhard Wessling)

Contents

How It All Began

Strange trumpeting calls rang out from somewhere in the bog directly in front of us. We had never heard anything like this before, and had no idea what it could be, but we were burning with interest. Their creators—they must be birds, but what kind?—were not, however, visible, because they were hidden behind bushes, trees and reeds. My then-wife and I, together with our two still very small sons, were exploring our new environment in that spring of 1982, because we had moved to Bargteheide only a few months earlier. The nature reserve "Duvenstedt Brook" was not far from our new home north of Hamburg, and we had already visited it a few times in winter, but this day marked the first time we were doing so in spring (Fig. 1).

We soon found out that these clear and powerful, far-carrying calls came from cranes, the first pair to establish a territory in Duvenstedt Brook in living memory, having settled there a year earlier. With their necks stretched high and their beaks erect, they trumpeted "oooo—i, i, i" several times in unison, over and over again—fascinating.[1] It was breathtakingly beautiful to watch the cranes during our increasingly numerous visits to the Brook: They danced around each other, swinging their wings, jumping up elegantly and springily, landing with the grace of ballet dancers and letting out short sounds as they danced.[2] Sometimes, they ended a dance with a unison call. Even when they were just walking across the meadows or through the bush, they did so calmly,

---

[1] cf. https://www.bernhard-wessling.com/Duettruf Rufaufnahmen: http://bit.ly/2Wem64C or http://bit.ly/2petlNJ (an overview of all the calls mentioned in the following notes with links can be found here: https://www.bernhard-wessling.com/Rufaufnahmen).

[2] Dance in winter time: cf. https://www.bernhard-wessling.com/wintertanz video: https://www.youtube.com/watch?v=qnOo2mB90-Y

Fig. 1   Wetland area in Duvenstedt Brook (© Bernhard Wessling)

Fig. 2   Common Cranes dancing (© Carsten Linde)

self-confidently, quietly searching for food here and there. But they could also gracefully walk around each other, looking at each other, or walk side by side with their heads up, presenting themselves to each other, either in preparation for a dance or instead of dancing. Simply beautiful (Fig. 2).

We were by far not the only ones who were captivated by these images. And almost every person who has seen migrating cranes in autumn or spring, who could hear their flight calls—over Frankfurt, over the Bergisches Land or Kassel notices and calls out: "Look, there are the cranes migrating again!" Meanwhile, more and more people are traveling to the Vorpommersche Boddenlandschaft in autumn to experience a magnificent natural phenomenon: Tens of thousands of cranes from Sweden, Finland, Estonia, Latvia, Poland and Ukraine rest here for a few days or weeks, flying early in the morning from their roosts in the shore zones of the Bodden to the surrounding fields and meadows to look for food. In the late afternoon or early evening, the flocks return, flying, accompanied by the rustle of their wings and their loud conversation, back to the huge reed beds to roost.

I was not able to experience all of this in my youth in the Ruhr region, nor as a young man after my doctorate, when I first worked in Duesseldorf, because there were no cranes. They did not fly over the Ruhr or the Rhineland, and the Bodden landscape would not have been accessible to me, even if I had known about it, because it was still in the territory of the GDR at that time.

An important step on my way to the cranes was a decision my then-wife and I made when our first son Bengt was born in 1978. We wanted to know more about nature. From the beginning, when we went for walks and hikes with our children, we didn't want to say "bird" (let alone "peep" or "tweety"), but rather "blackbird", "great tit" or "kestrel" whenever we would discover something and show it to the children. So, we bought a bird identification book and studied it.

Equally decisive was the fact that I took a new job in 1981 that required us to move from Duesseldorf to Bargteheide, a small town northeast of Hamburg. In the meantime, our second son Børge had been born, and we walked with the children, the youngest in his carriage and the older on our shoulders, to the nearby nature reserve, the Duvenstedt Brook, almost every weekend. I wanted to instill a love of nature and natural science in our children from the very beginning by observing and experiencing them together.

Shortly after we had heard and then observed the first cranes, we got to know some of the crane conservationists. The German Federation for the Protection of Birds (DBV, today: Naturschutzbund Deutschland, NABU) and the World Wide Fund for Nature (WWF) had initiated crane protection in Hamburg in 1982. Paths were closed, and visitors were informed and persuaded to be considerate in order that the brood would not be disturbed, not by humans at any rate. The concept gained greater success step by step.

At that time, I could not have guessed that, a few years later, I myself would be in charge of the conservation program. My predecessor introduced me to the subtleties of crane observation from the mid-1980s onwards, and ultimately handed this project over to me when he turned his attention to the cranes in eastern Germany and moved to Mecklenburg. I had already gained a lot of experience by that time, and had gotten to know the area and the cranes that were breeding there very well.

I was granted a unique opportunity to ponder crane behaviour in 1994, because of a pair that was breeding very close to a hiking trail. There was a nice waterhole, which had probably formerly been a pond for fish-breeding, but had now become marshy. Around it, bushes and trees provided nice camouflage. And the year before, this pair had settled there, laid eggs and incubated—until they were stolen by an egg thief. A shock for us! How could this nesting site, so well hidden, so difficult to see and almost inaccessible, have been discovered and robbed?

In Germany (and certainly in other countries as well), there are always illegal removals of crane eggs from clutches. During my time as the person responsible for crane protection, we counted three such cases—despite crane surveillance. The terrain is confusing, and at night, dawn and dusk, it is easy to hide.

In 1997, a bird breeder (who ran his business with state permission) was charged at Kiel District Court with, among other things, running his breeding operation with illegally procured eggs. He kept 60 cranes, including three foreign species. He had attracted attention by virtue of the fact that his crane breeding results were significantly better than those of the Walsrode Bird Park. A genetic analysis proved that the "offspring" were not related to their "parents", which led the prosecuors to conclude that he had obtained the eggs from wild clutches. During the investigation, a smuggling ring was busted in Mecklenburg that was stealing eggs from crane clutches and shipping them out, mostly to the Benelux countries. During one of these transports, more than 40 eggs were seized. The Schleswig-Holstein breeder was also found in the address file of the smuggling ring. Unfortunately, all of this did not convince the judge, who kept asking whether a witness had been present during the egg theft or when the eggs were handed over to the defendant, which, of course, was not the case. So, the breeder was not convicted, at least not for the eggs. Without question, however, there are breeding farms in Germany and Benelux, legal and possibly illegal ones, which breed and sell cranes from wild clutches.

There are also non-commercial motives for stealing eggs from the clutches of wild birds. In 1999, a ring of egg collectors was busted whose members exchanged eggs like other people swap postage stamps. What these people had gathered in eggs brings tears to the eyes of conservationists. Over 100,000 blown-out eggs were recovered, including numerous crane eggs, even some eggs of the Siberian Crane, of which there were, at that time, just twelve individuals left in the western population, and only one today.

A number of quite respectable people were arrested (not, as in the case of the trade in eggs to be incubated, some unemployed and homeless people from Mecklenburg who were hired by breeders). The hobby of collecting, blowing out and exchanging eggs of all bird species of the world was pursued by respectable customs officials, business clerks and geography teachers. The busted secret ring turned out to be only the tip of the iceberg (or tip of the egg mountain, if you prefer).

Back to the careless crane pair. In 1993, all attempts to safeguard its nest were unsuccessful. And a year later, the pair bred again in the same place. We discussed how to strengthen our guard duty, looking for places from which we could overlook the breeding area without being seen by visitors. Of course, there could be no real security. The thief (if he wanted more eggs) could have chosen times for further raids. Would it be better if we took the eggs ourselves, thus scaring the pair off and hopefully persuading it to re-breed in a more suitable place?

I was strictly against such an approach. The cranes had settled in this area and had to get along with the humans willy-nilly. If they couldn't manage here—under our guard—we couldn't help either. It could not be our job to constantly intervene when a clutch of eggs was inconveniently laid or endangered. So, we let the cranes breed. At least this time, the egg thieves should have a harder time, we swore, and we watched the surroundings of this nesting place very intensively. I and my older son Bengt, who was 16 years old at the time and with whom I once again spent an entire week on crane guard duty, checked the access points to the immediate vicinity of the nesting site every day at dawn and dusk.

Because of the nearby big city, one has a good view when the skies are cloudy above the Brook, even at night. Only a cloudless night with a new moon is reasonably dark, and on such a night, my son and I were once again guarding the crane's nest's vicinity at that time.

It was already late, and we actually wanted to leave. We were freezing, and it had become quiet. Bengt had come over to me from where he was standing. We stood in the dark, listening. The first thing we caught was not a sound, but a shadowy movement across the open field at the edge of the Brook, just inside the nature reserve. "Someone's there!" We crept forward and out of the cover of the site, so that we could see the person as he emerged from the next dip—if it was a person. Perhaps it would turn out to be merely a big dog, and the excitement would be all for nothing!

No, it was not a big dog, it was a figure completely dressed in black, who had even masked his face. We thought feverishly: What do we do? Wait until he goes to the nest? No, we might not recognize him in the undergrowth or even lose him. Besides, we didn't want to risk disturbing the nest. Even if this hooded man wasn't an egg thief, it was still within our purview to approach him, because leaving the trails is forbidden in this area, including at night.

So, we decided to run towards the person—Bengt on the outside, me on the inside—to cut him off. At this point, the masked man saw us, disappeared into the bend, did something that, as best as we could see in the dim light, looked like he was ditching a backpack, and proceeded hurriedly along the path. "Stop, what are you doing here?", I called to him. From what I could now see, he was a boy of about 15. He claimed to be playing his own cross-country game, but did acknowledge that he was off the trail in violation of the law. He denied any knowledge of a backpack, claiming not to have had one. Unfortunately, we couldn't find anything later in the darkness.

I stayed with the boy while Bengt ran to the nearest phone booth—mobile phones didn't exist yet—to summon the police. But they didn't come, so we finally called the official ranger. He was visibly uncomfortable; our perception was that he must have known the boy. The whole event ended like the Hornberger shoot-out: much ado about nothing. Whether the boy really wanted to steal eggs or whatever else was behind his behavior, we never found out. In any case, the eggs were not stolen that year.

But this did not relieve me of my concern for this crane pair, because cranes need three conditions for successful breeding: firstly, a constantly damp place with water about knee-deep. Secondly, absolute undisturbedness, because they leave the nest and/or the chick at any disturbance, making it an easy game for all kinds of predators, whether it be ravens or wild boars or foxes or martens. Without disturbance from humans, cranes do quite well with these predatory foragers. Third, cranes need an open meadow or clearing in a forest near the nest where they can walk and feed the chick after hatching[3] until they are fledged (which takes about 3 months). In the case of this crane pair, things did not look good in this last regard. Water was available. We made sure they were undisturbed as best we could. But where was the meadow or the clearing? Behind the breeding place (however about 2 meters higher, because the marshy pond with the nest lay below at a steep slope), there was a former pasture, but it was much too small and much too close to the main road, with people meandering along on sunny Sundays by the hundreds. No normal

---

[3] cf. https://www.bernhard-wessling.com/am-nest

Fig. 3 Crane pair on their nest in the pond, one chick has hatched (© Bernhard Wessling)

crane would endure this. On the other side, there was a dense forest with thick underbrush.

The day of hatching had come. I had found a spot from which, with good binoculars or telescopes, we could see almost directly into the nest without being seen ourselves. Never before or after did we have the opportunity to observe chicks at such an early stage. How small these two were, snugly wrapped in their golden brown "fur" and already quite alert! (Fig. 3).

Normally, crane parents leave the nest area with their chicks after a few days and walk to the meadow to feed them. The excursions become longer and longer day by day. At night, they either go back to the nesting site or look for another place (again, in about knee-deep water) to spend the night. When sleeping, they stand in the water, while the chicks sit in the nest. But these cranes hardly moved from their spot. They stayed in their little marsh, often less than 20 or 30 meters from the nest, or hid in the adjacent brush. All of this did not make a good impression on me, and I didn't have the slightest idea how the cranes would manage to raise (and fledge!) their offspring. Then, they grew up, and I came to realize that the place was very well chosen as a food source. It offered little exercise, but plenty of protein-rich food: The adult birds were able to catch many flying insects and feed them to their chicks.

Only, how was the rearing to continue when the juveniles left the confined nesting area? How were the parents going to show them what was good and what was less tasty, being so completely without an open feeding area? While, to the south, the nesting area was obscured and inaccessible thanks to dense bushes, a break-off edge (the nesting area was about two metres lower) and a narrow meadow from Hamburg's main access way into Duvenstedt Brook, to the east, north and west stretched almost impenetrable woodland with lots of undergrowth, including bushes under the birches and conifers. How sad that cranes are so tied to the breeding ground once used, and thus not at all flexible! That's what I thought at the time, when I saw the young cranes slowly growing bigger, because the parents were really skilled at catching the biggest insects. But how were the young ones to develop further as they grew and began their natural routine of wandering and getting to know larger and larger areas?

At the time, I had no way of knowing that the crane parents had a plan—more on that later. First, let's take a step back and look at what we crane guards, and crane experts and biologists in general, knew about cranes in those days. We can also look back a few centuries and millennia to see what cranes have meant to people over time.

Crane Knowledge Compact: The Myths and the Facts

Back then, as I anxiously watched the crane pair in its confined breeding site, apparently controlled by their purely mechanical instincts, I knew little more than a few basic facts about the birds' behavior.

Cranes and rails (like the well-known coots, which live on many types of water body, including park ponds in cities) are interrelated, as are bustards (which have been almost wiped out in Germany). These three families (and a few more) belong to the "Gruiformes" or "crane-like birds". Contrary to popular belief, cranes are not related to grey herons or storks. True, they look a little similar (long legs, long necks, long beaks), but that's all.

There are 15 crane species in the world, and these various species can be found on all continents, but not in South America, Antarctica nor in the most northern polar regions. However, their habitats are shrinking almost everywhere in the world. We will not describe the various species in detail, as only some of them will be mentioned in this book, and only four of them have been the subjects of my research. In Europe, there are two species (the Common or Eurasian Crane and the Demoiselle Crane), both of which can also be found in Asia, the continent where (in Eastern Asia specifically) we can find the highest number of species: eight. Many years ago, I had an extremely lucky day, when I went to the northeastern part of China, close to the border of the Chinese province of Inner Mongolia: that day, I managed to see seven different crane species, on that one day alone. There are a number of crane species in Africa and Australia (which I have not yet observed in the wild), and two species, the Sandhill Crane and the Whooping Crane, in North America. Apart from the Eurasian Crane, which I first got to know, I have also

Fig. 1 A Sandhill Crane pair with its freshly hatched chick (© International Crane Foundation, Tom Lynn)

intensively studied the Asian Red-Crowned Crane and the two North American species. These will be the main characters in later chapters (Fig. 1).

Cranes (i.e., the ancestors of today's species that are considered to belong to the crane family) have lived on earth for 60 million years. This means that they appeared 20–30 times earlier than some of the closer ancestors of humans (*Homo rudolfensis* and *Homo habilis*, about two million years ago). Crane ancestors have presumably been around since the dinosaurs disappeared from the earth. Of course, they have changed a lot during that long time, but they are one of the most successful species families in evolutionary history. There is growing evidence that cranes and all other birds are ultimately descended from dinosaurs—that they are actually modern dinos. Not only is the development of feathers in the evolution of dinosaurs becoming increasingly clear, the coloration of the eggs also points to a direct descent between the two: All bird eggs, regardless of their specific appearance, have only two color pigments. Those dinosaurs closely related to birds (the Maniraptora) also laid colored eggs, the coloring of which was based on the same pigments.[1] The Sandhill crane in particular seems to be closely related to the ancestors that lived about 60 million years ago.

The word "crane" has a very deep root in Indo-European languages, and it is obviously accepted among etymological experts that said root is

---

[1] cf. https://www.wissenschaft.de/erde-klima/schon-dinos-hatten-farbige-eier/ (German text).

onomatopoeic.[2] Its origin is *gérh2-no-* (likely translated as 'cry'), which is related to **g ar-* ('sound, call'), **g(w)erdh-* ('hear, sound'), and?**g(w) Rgh-* ('lament'). In Celtic, **gérh2-no-* is said to have transformed to *garano*. Thus, it is supposed that the Germanic originating **krana-* could be a Celtic loan word.

How widespread and commonplace cranes were in Europe in earlier centuries and millennia is shown in a study conducted by the Irishman Lorcan O'Toole.[3] Crane bones have often been found in archaeological excavations in Ireland—a country where there have been no cranes for the last hundreds of years. Many Irish place names are probably derived from the fact that they were former crane territories. O'Toole further suggests that place names of Gaelic origin in mainland Europe that begin with "Cor" or "Kor"—there are hundreds of those—date from the Bronze Age and are also related to crane sites. In modern Gaelic, there are a number of terms that may derive from earlier transferred meanings, such as "Corr duine" ("crane person") for "loner" or "Corrluach" ("crane portion"), meaning "leftover grain". The a. m. etymological article additionally mentions "corrguinecht", a form of "magical wounding" that involved adopting a crane-like posture while chanting satirical verse.

An Irish diptych (a two-part relief or painting with hinged doors that are opened to reveal the work) from 1399 shows the birth of Richard II: all of the surrounding angels have crane wings! Ancient tombstones and stelae similarly feature human figures with crane heads.

While in Ireland, as in England, the cranes have been wiped out by intensive hunting, among other things, the hunting of them has long been banned in Spain, Sweden and other European countries. This is certainly one of the reasons why the Common Crane population has recovered in some European countries since around 1980. Other significant reasons are the decades-long successful protection of cranes in the GDR, which led to "overpopulation", additional nature protection regulations, rewetted wetlands and the shift of crane migration routes westwards. However, the fact that there are more cranes again today may also be a result of the birds' ability to learn, as cranes are increasingly accepting breeding sites that they would have avoided in the past.

---

[2] The subsequent etymological information is taken from an article published by Piotr Gasiorowski, "Gruit grus: The Indo-European Names of the Crane", in: Studia Etymologica Cracoviensia, Vol 18, 2013, p. 51–68 (I thank Dr. Rainer Feistel for finding this article and sending it to me). I have copied the above information, including the symbols, from this article without an understanding of the meaning of said symbols.

[3] Lorcan O'Toole, presentation at the European Crane Conferenz 2018 in Arjuzanx/France (Abstract collection and personal notes as the proceedings have not been published yet).

Before reunification, cranes were almost extinct in the Federal Republic of Germany and had completely disappeared in Denmark and Norway. Only intensive nature conservation measures brought about a turnaround. In Germany, the crane has been removed from the Red List of endangered species. Before euphoria breaks out, however, it should be remembered that cranes used to be found all over Germany and Central Europe, even as far as Great Britain. In former times, long ago, there was notable distribution in France, Spain, Italy, Greece, Austria, Switzerland, Romania and Hungary—countries where cranes no longer breed today. In Germany, for example, there are still no breeding pairs in Hesse, North Rhine-Westphalia or Bavaria. Moreover, another formerly widespread Eurosiberian crane species, the Demoiselle Crane, can no longer be found in Middle Europe, not even in Rumania, and has long since disappeared from Italy, Spain and Portugal; it now only breeds in parts of Ukraine, southern Russia and Siberia, as well Mongolia and, to a small extent, in Northern China.

The Common Crane can now again be found in significant numbers in Sweden and Finland (altogether, over 30,000 breeding pairs and 100,000 individuals, respectively) and in Russia (around 20,000 pairs). After many years of few cranes being found in Norway, the population has now risen again to almost 5000 pairs. In small and densely populated Denmark, there are now almost 500 pairs. In Great Britain, surprisingly, three cranes arrived in Norfolk in 1979; a first breeding attempt started in 1982, only one year later than in Hamburg's Duvenstedt Brook, but was not sustainable. The reintroduction of cranes to England was later supported by eggs from wild clutches carefully selected in a very crane-rich area in Brandenburg (East Germany). These were then artificially incubated, and the cranes had been released into a group of already wild cranes in the area. The first breeding after hundreds of years in Ireland happened just recently in 2021! In the meantime (after ongoing reintroduction of artificially incubated cranes in southwest England), a little more than 50 pairs are breeding more or less regularly in England. Another crane pair bred two or three times in Normandy, but did not stay there. More recently, there have been a handful of pairs in Lorraine.

In Germany, the crane population has recovered considerably: it is estimated that there are about 7500 breeding pairs, the vast majority of which live in Brandenburg and Mecklenburg-Vorpommern. Smaller occurrences can be found in Poland, the Baltic States and the other eastern successor states of the USSR. Mongolia, Kazakhstan and China also have a few hundred pairs each. In addition, there are about 300 breeding pairs in Turkey. The *Grus grus* species is said to have four subspecies,[4] the major eastern one being somewhat lighter in colour and smaller in size.

Estimates for Eurasia (Europe, Siberia, Mongolia, and north-east China, with wintering in India, for example) as a whole are about 450,000, perhaps as many as 500,000 individuals.

Sweden, where there really is enough room for people as well as cranes, allows farmers to shoot cranes if they enter their fields too often. Our Swedish crane conservation friends apologised for this at the European Crane Conference in Stralsund. Particularly in the resting and wintering areas, the conflicts between cranes and farmers, between nature conservation and agriculture, although very localised, cannot be overlooked, as is also increasingly the case in Germany.

A female crane weighs a modest 4.5–5.8 kg, despite its imposing size (cranes grow to 1.10–1.30 m tall, with a wingspan of 2.20–2.40 m). Male cranes weigh 5.1–6.1 kg. By comparison, the much smaller greylag goose weighs around 3.5 kg, and a mute swan up to 12 kg. To maintain its body weight, the crane needs surprisingly little food. Its stomach normally holds about 100 g, of which 30 g alone are small stones. However, when thousands of cranes forage on a field, quite a lot is gathered, despite the small amount of food required per individual, especially as a group of cranes in spring on a field with, for example, freshly germinated broad beans can completely ruin said field if they are not scared away, because they uproot the plants. Crane protection, like nature conservation and species protection in general, must also face the conflicts of interest with agriculture, among others, and help to resolve them creatively.

Cranes eat roots, tubers and bulbs of various plants, potatoes, seeds, corn and cereal grains. As for animal protein, they eat larger insects and their larvae (crickets, butterflies, dragonflies, beetles), earthworms, and occasionally lizards, frogs or mice. However, unlike herons or storks, they are not good

---

[4] Most experts say that there are four Eurasian Crane subspecies, one of which is the endangered subspecies "*Grus grus archibaldi*" in Caucasus. Subspecies are only vaguely defined by morphological differences, and a definition is even less clearly available than for the question "What is a species?" Currently, it is the generally accepted opinion that there are *G.g. grus* (western population), *G. g. lilfordi* (eastern population, mainly in Russia), *G g. archibaldi* (Caucasus) and *G. g. korelovi* (Tibet, Kazakhstan, Kirgiztan), cf. V. Y. Ilyashenko, Proceedings of the European Crane Conference 2012 (Stralsund), p. 117 (2013). However, any genetically convincing differentiation is still missing. A genetic distance of 0.1 to 0.3% was found by M. Haase and V. Ilyashenko (Andreola **59** (1), 131–136, 2012), "hardly supporting the distinction of the morphologically defined subspecies", as they wrote; they had evaluated samples from *archibaldi*, *lilfordi* and *korelovi* and compared them with *G. g. grus*. Similar results ("low genetic differentiation") had been published by E. Mudrik, T. Kashtentseva, P. Redchuk, D. Politov Mol Biol. **49**, (2), p. 260–266 (2015). The subspecies *grus* and *lilfordi* are showing a vast morphological continuum; *korelovi* seems to be morphologically more distinct, while *archibaldi* is clearly differentiated.

hunters. Nevertheless, I once happened to observe a crane catch a small bird that was flying by and immediately devour it. It is generally said that cranes do not feed predatorily. However, this view seems to be gradually changing at present, due to occasional observations—which are anything but easy to make. For a story on this, see the appendix under the headline "Peaceful Cranes?"

A crane takes in an average of about 130 g of food a day, and about 180 g in the wintering areas. It is therefore astonishing that cranes are such persistent flyers. On the way to their wintering grounds, they can apparently fly for 24 h at a stretch (perhaps interrupted by short breaks), but on average, they tend to fly only 50–200 km a day. During migration, cranes spend about three quarters of the time in resting areas. The total flight distance—depending on the breeding area and the location of the wintering grounds—is up to 4000 or even 5000 km. Their flight performance is astonishing, because cranes are so-called rudder fliers, i.e., in contrast to storks, they are not mainly carried in gliding flight. Therefore, their speed is not particularly exciting: depending on the wind direction, they reach an average of around 45 km per hour.

As a rule, cranes do not fly higher than 1000 m, but pilots have also encountered them at altitudes of 4300 m. Black-Necked Cranes living in Tibet or in the Tibetan highlands in western China have even been observed at altitudes of over 8000 m on their way to their wintering grounds south of the Himalayas. The same is partially true for Demoiselle Cranes. Flock sizes vary during migration: On departure, groups are around 100 birds strong, decreasing to 60–70 towards the middle of the flight, then to 25 towards the end of the route, until finally only around ten individuals form a flock. For short periods, 400–500 cranes can be found flying together, but this is very rare.

The trumpet-like calls of the Common Cranes on the ground are unmistakable, mostly made in pairs in a duet. They can sometimes be heard for miles. This volume is possible due to their up to 1.30 m-long trachea.

Cranes have very good hearing, probably not the best sense of smell, but excellent vision. They can detect any suspicious movement, even from a distance. If you try to approach a crane (which you should avoid by all means, because we don't want to disturb wild animals, but photographers and researchers occasionally have to try), you will usually find that it has already seen you coming long before you have seen it. People used to say, "Cranes have a hundred eyes on their tail feathers," because the birds recognize us humans even when their backs seem to be turned towards us.

Memory and the differentiation of images also seem to be very strong in cranes. They can tell us humans apart, while we struggle to distinguish the "face" of one crane from another. (Truly, they all look the same to us—despite that apparently not being the case in the other direction; you can read about this in the appendix under "USA: human friends and enemies".)

Cranes are sexually mature from roughly the age of four. They can live about 25 years (on average, reportedly 13 years), or over 40 years in the zoo; in regard to the Siberian and Red-Crowned Cranes, it is reported that (in captivity) even 80 years are possible. I have myself been able to confirm a high age (at least 26 years) of a wild Whooping Crane, the "Lobstick" male, about which I will tell you more in chapter "The Adventure Continues: Expeditions to the Wild Whooping Cranes". The oldest captive Whooping Crane lived to 46 years old.[5]

The scientific literature on cranes[6] cannot be compared with that, for example, on great apes, neither in volume nor content—incomparably more research and publication is done on primates and other mammals. The crane literature from around the world includes several thousand sources and is dominated by work on crane distribution: how many are found where and when, changes in breeding, migration and wintering numbers, research on migration routes. The next largest group consists of works on the nature and quality of breeding, feeding and wintering grounds, and the food itself.

If one searches for literature on the behaviour of cranes, hardly anything can be found. The few publications that exist consist of purely systematic (sometimes even statistical) evaluations of observed movements of individual cranes at the nest or while feeding ("flies up, stands, preens, pecks, secures ..."). Courtship and dancing, territorial defence, breeding and rearing are described in detail. It is remarkable that so many have been tempted towards a mechanical description and assigning the observed behavioural elements to "functional circles". These include "posture", "sleeping/resting", "paying attention, watching", "grooming and comfort behaviour", "locomotion" (i.e., "movement from one place to another"), "foraging and food intake", "antagonistic behaviour" (i.e., behaviour towards other cranes) and "reproduction, courtship".

The enumeration of 132 different postures (as, for example, for the Japanese Red-Crowned Crane) or of about 50 postures in the eight "functional circles" gives the impression that the "behavioural inventory" of machine-like animals is described.

---

[5] https://www.msn.com/en-us/news/us/first-whooping-crane-hatched-at-baraboo-foundation-dies/ar-BB1epvfj

[6] A good overview of the literature on cranes can be obtained from the "International Crane Foundation" with a computer-assisted search. The library and database there should be almost complete.

All of this work is important, pertinent and helpful, but does not go into any depth. After all, what do we really know about cranes if all we know is their "functional circles"? I mean, terms like "functional circle" and "behavioral inventory" mislead us into overlooking an important competency: the ability to behave in flexible ways that are appropriate to the situation and environment. I am not aware of a single publication that addresses the actual behavior of cranes, that is, their social behavior, their responses to the environment, and their intelligence and ability to learn, especially in the field. Field research projects like those that have been done on many animal species (mainly primates and other mammals, less frequently on birds, such as ravens) have never been done on cranes.

Behavioural studies on birds are usually carried out in captivity. Here, grey geese, pigeons, parrots and corvids are popular study objects. However, despite the fact that many cranes, including many Sandhill Cranes, are kept in two facilities in the USA, there was and is no behavioural research even there. So, we are not much further along in terms of crane behavior and cognitive abilities than humans were in previous centuries and millennia.

In fact, numerous myths exist around the crane. Especially in Asia (China and Japan), the superstition that cranes live very long lives, perhaps that they are even immortal, was or is widespread. This is due to the fact that, where people know crane territories, they have observed over generations that "every year the same cranes" bred there. Until only recently, there was no way to determine which crane individuals were breeding there this year, then the next, and finally in the following years and decades.

In the Chinese language, there is the word "鹤年 "(pronounced "hè nián", where the accents indicate the correct stress of the syllable); literally translated syllable by syllable as "crane year(s)", it means "long life", and some parents name their child accordingly in the hope that it may have a long life full of enjoyable accomplishments.

In Japan, but especially in China, there are countless paintings and silk embroideries showing cranes on pine trees, which are typical there.[7] I lived in China for 13 years, and I often asked my friends there if they did not know that cranes (unlike herons) could not land or stand on the branches of any trees. Their talons just aren't up to it. But none of the Chinese people I know (and that's a large number) had ever seen cranes, let alone were able to tell cranes and herons apart. So, I came to believe that the artistic representation indicated ignorance on the part of the painters and a confusion with herons, until I learned that both this species of pine on the one hand and the cranes

---

[7] As just one of countless examples: https://bit.ly/2G7PnIk (last visited Sept 22, 2021).

on the other both stand for "long life", with the pines additionally representing robustness—so both together express, in duality, the wish for a long life.

While the crane is highly respected, at least in Asia, it is portrayed in a fable by Aesop as naive and stupid, as a loser: A wolf has all too greedily eaten a sheep, and thus a bone has gotten stuck in its throat; a passing crane agrees to help it, sticking its head in the wolf's mouth, pulling out the bone with its beak, and then demanding the promised reward; but the wolf rejects this, saying that it is reward enough that the crane was allowed to pull its head out of its mouth unharmed. This fable probably reflects a particular opinion about cranes, which were considered elegant and virtuous, but not "clever" like wolves and foxes.

In Japan, on the other hand, the crane stands for authority. If you want to say that someone's word carries weight and does not tolerate contradiction, then this is "the word of the crane".

Already among the ancient Greeks, the crane had developed into a symbol of vigilance: it was believed that it slept with a stone in one of its two claws, which it held lifted high; as soon as the stone fell out because it had fallen deeply asleep, it would wake up. Also, cranes do sleep standing on one leg, like flamingos, but are not often observed doing so. This is due to the fact that, unlike flamingos, cranes can only very rarely be observed at their roosts. Nevertheless, one can find drawings and figures on old houses that indicate this ancient belief—by the way, this includes Germany. In this country, the Mendelssohn Society uses the watchful crane with the stone in its claw as its logo, along with the motto "I guard". The German writer Wilhelm Busch (1832–1908) wrote the endearing poem "Last but not least, the wise crane" about it.

Ignorance and pure fantasy also form the background for the strange interpretation of the crane migration to the south, as found in Homer's Iliad: according to this, man-eating cranes fly south to fight with the pygmies in Africa. The migration of cranes also plays a role in Schiller's ballad "The Cranes of Ibykus", as they are the only witnesses to a murder; when the murderer sees migrating cranes once again, he exposes himself.

In China, the crane is often the protagonist of stories, fables upon which some of the thousands of beautiful and wise Chinese proverbs are based. In one of these tales, a fox invites a crane to dinner, but he serves the soup on a flat plate so that the crane cannot drink it. The fox simply slurps it up with his tongue. Soon afterwards, the crane sends him a return invitation, but this time, the meal is to be found at the bottom of a jug, which has a long, narrow neck, a bottleneck. The fox thus realizes that his trickery has resulted in his being tricked in return, for he cannot reach the bottom of the jug. This story exists in many variants; in Europe, it has been known since ancient times as one of Aesop's fables, sometimes with changing actors, for example, a stork instead of the crane. But it seems to have its origin in China.

In another millennia-old Chinese story, homage is paid to a general who, when his kingdom is invaded, stays with his king despite enormous losses, protects him, and saves the country. The people at the time said that he was as outstanding "as a crane standing among chickens." The Chinese have made proverbs of this, as they have with thousands of other stories, by condensing the story into exactly four or eight characters, and every reasonably literate person understands the background. In this case, the corresponding "Cheng Yu" (which is the name given to this kind of proverb) is "鹤立鸡群 hè lì jī qún", literally "Crane stands (in) flock of chickens". "骑鹤上扬, qí hè shàngyáng", which would be translated as "Riding (a) crane to Yang(Zhou)", in which YangZhou is abbreviated to "Yang" alone and denotes a town in East China. One would use this Cheng Yu to indicate that someone has reached a higher, more attractive position. "风声鹤唳 fēng shēng hè lì", "sound of the wind (and) a crane call", denotes particularly fearful people (originally soldiers, but nowadays any person) who immediately panic at any unexpected sound.

In all countries of the world where there are cranes, they are considered lucky charms and messengers of spring, because they return in the spring and then dance so beautifully. "Spring messenger" is a perfectly a propos designation from a purely technical point of view: When the cranes migrate back north in February, when we hear their flight calls at night, it will be spring in the foreseeable future. "Lucky charm"—well, that's up to each person to decide and is beyond scientific analysis. Cranes have brought me a lot of luck: whenever I see them, and this still remains true today, a smile flits across my face and my mood lifts, a lucky feeling indeed.

According to one of the oldest and most persistent myths, crane pairs are faithful to each other for life; i.e., it is believed that a crane has only one partner in its lifetime. This view is particularly popular in China and Japan (although most people there have never seen a crane). In Japan, cranes are therefore extremely popular as an image on wedding dresses. After all, this idea is all too beautiful and so comfortingly romantic. This belief, which is still widespread among crane enthusiasts, experts and researchers, will be addressed later. In the next chapter, I would like to devote myself once again to the crane pair that was stubborn enough to breed anew in a place that we considered to be unsuitable.

Problem Solving, Ballet Courtship and the Fox Alarm: How Do Cranes Communicate with Each Other?

The crane pair that again nested where they had been robbed of their eggs the previous year worried me more and more, because there was no meadow at all in the immediate vicinity on which they could guide and feed their chicks. And for weeks, they did not show any sign of an intention to leave this small marshy pond, which was absolutely unusual for cranes as I knew them.

One sunny Sunday morning, I was again able to observe the crane parents with the chicks on the marsh near the nest. The chicks had grown nicely, had been fed very well by the parents, but had no space to start walking and exploring the world. By this time, I had almost gotten used to the idea that they would probably never get out of their self-chosen bog hole, and thus the brood would be unsuccessful. Meanwhile, 3 weeks after hatching, the young were already the size of chickens or turkeys.[1] We had never before observed a crane family staying so long in the immediate vicinity of their nest.

On the afternoon of the same day, I again went to the Brook. The crane family had disappeared. It had already happened several times that I had not been able to find them or could only do so with difficulty, when they had hidden themselves—for whatever reason—particularly well. But this time, the birds were really out of sight; although I patiently surveyed the area from various angles, I could not find any of the four cranes. I was worried.

Out of pure routine, I stopped here and there on the way home, searched some other places with the binoculars, and suddenly could breathe a sigh of relief: The family I was looking for was in a meadow 500 m away from the breeding site and separated from it by a rather wild, shrubby and barely passable forest. It was clear: the cranes must have set off around midday (I had last

---

[1] cf. https://www.bernhard-wessling.com/youngster

seen them around noon) and had arrived at the new meadow by 5 p.m. at the latest (that was when I discovered them there), probably much earlier.

I was speechless. For weeks, I had been thinking about what the crane parents would do to solve their problem, and it turned out to be so simple: They waited until their offspring became big and strong enough to walk with them to the meadow, climbing over branches and fallen trees and through the bushes.

In spring, I had observed a crane pair on this very meadow during courtship—now, I assumed that it was the young parents that I had here again before me. This was "their" meadow, according to them, it belonged to their territory, and it had clearly been "obvious" to the pair from the beginning how they would raise their chicks, when they would change location, etc. The adult birds had probably never "considered" using the meadow immediately adjacent to the breeding territory, right besides the main hiking trail. The number of people who were there at times had certainly not escaped their notice.[2]

I was impressed. I had been worried and had not thought of the obvious. It was now evident to me that this pair had developed a plan for providing food for their young since their arrival in the territory after wintering. They had also previously chosen a roosting place close to the open meadow (again, a sufficiently moist place), which was unsuitable as a nesting place but ok for roosting—because this is what they were now aiming for every evening.

My experience gave me food for thought: I had thought the cranes were not particularly intelligent, or at least more stupid than they had presented themselves to me here. I had certainly overlooked many an indication in the years earlier of how intelligent cranes are, how flexibly they adapt to their environment and surroundings and manage to solve problems and hold their own even under adverse circumstances.

Ashamedly, I had to admit to myself that I had succumbed to human hubris: We always believe that our extraordinary feat of intelligence is that we manage to defy the adversities of nature. We build ourselves tools, houses, and cars, and thus circumvent our natural limitations.

---

[2] I conclude this from my numerous observations over the years, according to which cranes seem to know not to use foraging grounds close to the hiking trails on Sundays when the weather is good. On a sunny and warm Easter Sunday or May First Holiday, you are not likely to find any crane anywhere in the Brook; they will all have retired into the woods and behind the reeds, because so many people are there. On weekdays in early morning, one can find them very close to trails; also on a Saturday or Sunday, but only if the weather is "bad": then I may be the only visitor to the Brook for hours; and I have often met a crane pair or a crane family quite close to a trail, not only in my earlier years as a crane guard, but nowadays as well. It was in this way that I once happened to take a photo of a crane family with my mobile phone, because I had not brought my good big camera with me owing to the rainy weather. But while the roughly 6-week-old juvenile was only a few meters away from the trail, one of the adult birds another ten meters away was trying to lure the offspring back into the foliage, away from us, with no success.

The abilities of animals to survive in their natural environment (or in the environment changed by us humans), or even more, to live a rich, challenging, eventful and beautiful life according to them (and not just to fight daily for survival, as it is often so beautifully creepily put), we then call "inherited". We think that animals can do this "automatically", because their instincts have programmed them to do so. We overlook how much intelligence and ingenuity lies behind the behaviour of animals and their adaptations, and how many of their successful behaviours they had to learn, laboriously, but cleverly.

For me, this event was a key experience. I resolved that my observations would be even more intensive, but, above all, different. More impartial. Since then, quietly looking at, listening to, smelling and feeling the processes in nature has made me downright happy countless times. Time and again, I have had and continue to have opportunities to reflect on the amazing feats of evolution—for us humans and for all other living creatures. This has made me more and more humble during the course of my life—more humble about my role as a human being in the world and about the role of humans in nature.

I have taken even more time since that experience for watching cranes (and not just them!) at various points on the Brook, at all times of the day and year and in all weather conditions. As I frequently walked (or biked) there alone or with family, I didn't even feel it was an extra time commitment, but rather a welcome exercise and distraction from my very strenuous but equally interesting job as a research chemist and entrepreneur. Additionally, I read books and articles, and talked to people who already knew cranes better than I did.

In spring, it is particularly exciting. When the cranes have found their "home" in their own territory and breeding area, the partners become more active. They mate, they let us hear their unison calls, they strut and dance. The days becoming longer in spring, in concert with the courtship itself, causes eggs and sperm to develop until both are ultimately mature and can be united after copulation. It is characteristic of cranes that it takes weeks of (rather cumbersome) attunement to achieve fertile mating—otherwise, sperm would meet incompletely developed eggs, or if the female crane laid her eggs, they would not be fertilized.

As a long-time crane guardian, one often has the opportunity to observe courtship dances and copulations—provided one does not shy away from the early hour before dawn and the usually long wait, quite often with no remarkable observations. I don't remember how many times I have seen courtship and copulation, but I was certainly able to observe each of "our" pairs several times. (And I have even less of a memory for the far greater number of times I observed nothing.)

In many cases, but not always, mating is initiated by a courtship dance. The two cranes walk slowly around each other, in a "measured" way; sometimes the female appears uninterested, and the male tries to attract her attention. As they dance, the birds leap a quarter meter high (or even a whole meter!), then to the side and forward. As they do so, they flap their wings vigorously, but not frantically, duck jerkily, and raise their necks again. There is no fixed choreography, no hereditary ritual. Every couple dances differently each time.

Before laying eggs, cranes copulate frequently, mostly in the morning, not long after waking up. The early hour is thus the most unpleasant thing for the one seeking the highest probability of observing them engaging in this activity. One gains some time after the change to daylight savings, and so, around about the middle of April, one must "only" get up at three o'clock in the morning if one wants to be there at least an hour before the cranes and still hear oddballs, spotted rails, snipe and woodcock, or perhaps see a mighty wild boar, or at least hear or even smell one in the immediate vicinity, and this with at least some breakfast in the stomach.

Sometimes directly after the first call, sometimes an hour later, sometimes without a previous unison call (and, much too often, not at all), the crane pair appears on the preferred dance floor (with exceptions, they always seem to be courting on the same, usually particularly difficult-to-see area; but I have to admit that I recently also once had the chance to watch a courtship and subsequent copulation in plain daylight from my terrace: the closest crane territory is only about 500 m away from my new home where I now live with my life partner; this crane pair can often be seen from our home when it visits a meadow in front of the house; on this special day, I could see the cranes would start mating soon, so I called my partner to come out immediately to watch it—and yes, indeed, they delivered, presenting her first opportunity to see copulating cranes) (Fig. 1).

Back to the observations from past years. The two partners stride around each other, dancing. This can drag on for a long time or last only a few minutes. Finally, the male jumps onto the female from behind, sits astride his mate, somehow holds on to the female's wing joints with his claws, and mates with her within a few seconds, flapping his wings (to keep from falling off).[3] As they do this, you can hear a noise that sounds like a groan or moan—but this is only if, like me, you had the opportunity to be there in time to get relatively close to the mating site before the cranes do and wait patiently. After copulation, the male jumps off over the female's head. Often, courtship is over immediately after this, the female arranges her plumage, and both partners go

---

[3] cf. https://www.bernhard-wessling.com/kopulation

Fig. 1 Common Crane pair copulating (© Carsten Linde)

about looking for food. Sometimes, they will again let a loud and cheerful unison call resound.

One morning during my one-week shift as a crane guard, I experienced the most beautiful courtship display of my "career". I had cycled in deep darkness to a crane territory and walked from the path farther into the area, climbing a small high perch in the middle of the crane territory. From there, I wanted to determine whether a crane pair, which I called the "Sly Dogs" (I'll tell you why later), were preparing their brood and courting, whether a breeding attempt aborted 10 days earlier would make this season a failure for the pair, or whether they would try breeding again. (I was authorized to leave the public paths and enter into the otherwise non-accessible inner areas because of my official role as a nature conservation guard.)

Rain was forecast, but when I set off on my bike from the forestry depot (where we were staying overnight in the center of the Brook), it was just drizzling. When I reached my raised hide, it was windy and still dark. I heard a Spotted Crake calling and two Snipe making their territorial flights, producing characteristic sounds with their tail feathers (you might think the Snipes are "calling", but no: their tail feathers vibrate and produce these sounds).

Only a few minutes later, the wind freshened up, developing into a storm and bringing slapping rain with it. Despite good clothes that were supposed to defy the rain, I was already chilled to my bones after an hour. I considered for a moment going back to the warm room, but decided not to admit defeat. So, I persevered, further motivated by the dawn, which promised some clearing of the rain clouds (which changed from dark grey to light grey). And sure enough, another half hour later (I was pretty frozen stiff by then), the storm subsided, the day became bright (grey)—and the cranes called, a full hour later than usual.

This reassured me, as it showed me once again that they too are sensitive to the weather and do not greet a rainy, stormy morning joyfully, including in their protected swamp forest. I found myself in an even more positive mood when "our" white-tailed eagle flew over me. He had probably spent the night in the Brook, perhaps even after catching a heron in a wild chase the day before (according to Bengt's observation).

Ten minutes later, the "Sly Dogs" stepped out onto the meadow where I expected to see them, not far from me. The now only light wind blew across from them right in my direction.

And then, the dance began. I was happy to have waited after all. They both danced, after which he jumped on her, I heard the moaning groan during copulation, he jumped off—and danced on! Now even more beautifully and intensely than before. The jumps went farther, while she walked at measured steps behind him.

But then, he stopped all at once, turned his back to her, unfolded his wings, and looked to the side, practically representing the German federal eagle symbol, the Coat of Arms of Germany.[4] He did not spread his wings completely in full wingspan, but let the wingtips hang down. The light gray wing tops and black arm wings made a glorious contrast; the decorative back feathers were erect. Like a peacock, he presented himself, only more impressively, with wings spread and arm wings hanging down, head raised high, turned to the side. The female made her neck long, strutting around him with her beak pointing upwards some 5 meters away, as he constantly showed her his backside, thus turning as his partner passed behind him, in slow motion, so that she could always see him in full breadth and beauty.

Yes, he was a beautiful and strong male crane, without any doubt.

It was the most fascinating courtship I was ever allowed to experience. The coffee afterwards tasted twice as good. This experience also gave me food for thought. I had often read that the courtship display of cranes, with their dance, is genetically determined. That is, the animals could not help but behave as it corresponds to "the norm". How does it come to these deviations then? Every dance I had observed so far was already different from other dances, every courtship display was different from what I had ever observed before. More so this one, in which the male resembled a peacock or a striding

---

[4] https://en.wikipedia.org/wiki/Coat_of_arms_of_Germany

federal eagle. I had seen a few courtship dances from this pair, as well as from others, but never anything like this early morning.

How does this flexibility in behaviour come about? How does the crane suddenly come up with such a courtship element? Will he do it again? Could it have been pure coincidence that he turned exactly in such a way that his partner could always look at him in full breadth? Or does it not rather show us that he himself has an idea of what it looks like from his female partner's point of view when he poses and moves like that? Doesn't this show us the crane's ability to put himself in the place of another individual, that is, to be able to imagine what "the other" can see or perceive? This is called "Theory of Mind" in behavioral research, and at the end of this book, I will discuss this topic in a little more detail.

With each new observation, things became more interesting. Soon, I wanted to know who exactly I was looking at. We didn't give "our" cranes names, as one would do with pets. In the first few years of crane watching, no one could definitely *know* if it was the same pair breeding in the same place as the year before; we just assumed so, because that was what the books told us to expect. Crane experts said and wrote that crane pairs always return to their ancestral place. I had no reason *not* to believe this as well, and was by now convinced that we were dealing with the same cranes over and over again. However, there were also remarkable changes that manifested as early as the following year, and did so even more over the years.

Visually, we can not distinguish the cranes. We can see that one of the partners is usually somewhat larger than the other (thus, one can assume that it is probably the male; it is also in our nature to believe this, although there are exceptions, and with hawks and sparrowhawks, it is actually the other way around).

The red head-plate[5] changes in extent and brilliance too rapidly during the spring to be used as an identifying mark. Although the experts in the books write that males and females have red head-plates of different shapes, and we notice in some that they are of different sizes, the cranes are often so distant that determining the head-plate, which changes anyway, or the size leads nowhere. Plumage colour also changes, although it is distinguished by the fact that, interestingly, it is actively changed (see chapter "What Can We Learn About Intelligence, Migratory Behavior, the Formation of Culture, Tool Use, and Self-Awareness in Cranes?").

So, for years, I had only the behavior to distinguish the birds, to recognize single individuals. I had gotten into the habit of paying attention to the cranes'

---

[5] cf. https://www.bernhard-wessling.com/Kopf-Nahaufnahme

daily behaviour patterns. For example: Where does the crane or crane pair go in search of food, where do they feed before mating or before breeding begins, how far can you approach them before they flee, where does one partner go in search of food when the other one is sitting on the nest, from which direction, which trajectory do they use to fly into the nest area, where do they guide their chicks out, etc. These habits vary from pair to pair, from individual to individual and can be used to distinguish individual pairs from others, as well as within a pair to distinguish the two partners from each other.

I thought I could tell from the behavioral pattern whether it was the same pair as the previous year occupying a territory, or whether there had been a change. I was not so sure if it was always the same mate that the (male) "territory owner" (or the female territory owner?) showed up with the next year.

We crane guards, of course, chatted about the bird pairs. We made accurate records of each day's observations, using the names of the territories to identify the pairs. Anyway, we didn't call them "Klara" or "Else", "Karl", "Timotheus" or "Cleopatra" (which some crane watchers and researchers prefer), let alone "2GX13B" (although that would look very impressive, of course).

In this book, however, I would like to deviate from this practice and give the crane pairs names to distinguish them. Since I do not want to use the territory names, I have come up with names that characterize their behavior. Thus, I called the supposedly stupid and stubborn pair, which had proved me wrong in this assessment, the "Planners", and in the following, I relate an incident with the pair dubbed the "Pioneers".

I call this pair the "Pioneers" because they bred at the site first chosen by a crane pair as a breeding site in the Brook in 1981 and 1982. Since then, this site has been used without interruption—mostly successfully. There was a high probability that the pair that arrived here in 1981 was not the same pair that was still breeding here in the mid/late 1990s—those would be truly senior cranes! (Theoretically possible, but not too probable.) But it's amazing that there hasn't been a single year when this breeding site has gone unoccupied. What we were most likely dealing with here was a "continuity in change".

Anyway, our "Pioneers" were once again breeding in their regular spot. But in the year in question, almost all pairs broke off breeding at about the same time, because the water level fell by 15 centimeters in the whole brook within a week—a problem that has to do with the still intensive drainage and that always occurs when a lot of rain has fallen earlier. As a result, there is a high water level, and if a period of a few days with no rain at all follows, the water level quickly falls, with the result that nests will no longer be protected by surrounding water and will be easily accessible for foxes and wild boar.

The other pairs soon began another breeding attempt, which was not unusual. The "Pioneers", however, took their time, mating so that the noise from their activity resounded throughout the brook, but did not result in the laying of eggs. Finally, the time had come. It was the fifth of May, which is really late. So, in early June, the chicks hatched, and a few days later, the family appeared as expected in the open meadow to feed the youngsters.

Since it was already so late in the year, the grass had grown accordingly high. Even if the chicks hatch and are guided out a good 6 weeks earlier—as is usually the case—it is hard enough to spot them in the grass in the first few days because they are so small. Now, however, the grass was even taller. We stared into our binoculars until we got headaches, but whatever we did, we didn't get to see more than a small brown tuft. Impossible to determine if it was one chick or two (Fig. 2).

Well, then, we just had to wait a few days until they became big enough. Sounds reasonable, except that the grass was growing too, and it was growing faster than the crane chicks. We were still unsure if there were two or just one, although there were signs indicating two of them.

We also became more and more restless because we noticed that the adult birds seemed uneasy. All too often, a little one seemed to have disappeared in

Fig. 2 A crane pair accompanies their chicks in the grass, but is keeping away from the higher grass in the background; if the offspring is hatching early enough, it will always be taller than the grass, but not if it hatches much later. (© Bernhard Wessling)

the tall grass, so that the parents had to search for them. The stronger the chicks became, the more cheeky (and age-appropriately careless) they also became. Even in a meadow with grass that is not too high, it is interesting to observe how sometimes the one, sometimes the other juvenile goes its own way and escapes the supervision of the parents.

Now, the little ones knew nothing but tall grass and obviously felt comfortable in the grass forest. But the adults were uncomfortable, that was clearly noticeable. Us too, because, just like the parent birds, we had not failed to notice that the resident fox often crept up through the grass (Fig. 3).

Actually, he was more interested in the grey geese and their chicks in a neighbouring meadow (there were simply many more of them compared to the mere two crane chicks), but the feeding meadow of the "Pioneers" was on the way from his burrow to the geese, and such a crane chick brings a little rare bit of variety into the menu. This is a drama that we have borne witness to a few times. If the cranes notice him in time, all is well, for they drive him away effectively—the fox has respect for the long beaks and the huge birds with the wide-open wings. Perhaps he has already felt the beaks and does not want to experience that again. But creeping up in the tall grass increases his chances!

Fig. 3 Also this pair has no problem anymore with their offspring to look for food in the tall grass of this clearing. As this juvenile had hatched in May, at even not 2 months of age, it is already almost as tall as its parents. If it were only 3 or 4 weeks old, the grass would be too high for the adults to keep track of its own expeditions into the grass forest. (© Bernhard Wessling)

The question for me was whether the cranes would surrender to the problem without resistance. They had to be vigilant as hell, because the fox could sneak up unseen from any side at any time. We often saw him sooner than the cranes themselves (and cranes really are extremely observant, no movement otherwise escapes them). And then there were always those cheeky youngsters, who never wanted to stay with their parents who care for them so much!

One day at noon—suddenly and, from the crane guards' point of view, without any advance signs that something new would happen now—the adult birds marched off, up the slight slope of the meadow, with the chicks following. In the middle of the day, they crossed a hiking trail—only a few dozen meters away from the human observers (two crane guards, two hikers)—and hit the bushes of the adjacent bog birch forest.

But one of the chicks had gotten lost, probably scared because of suddenly being near people. At the same time, the farmer who had leased the meadows and used them (extensively) had appeared all of a sudden. He had followed our crane protection very closely and liked to talk to us about his and our observations, about progress or failures. Now, he saw that the one chick had fallen behind the other three family members, seemingly irritated by the people who were in the way, and was fleeing back towards the familiar meadow.

The farmer had noticed that the parent cranes had disappeared into the bushes with the other chick, as well as the fact that our crane guards were too far from the lost son. Without further ado, he walked a few long steps behind the poor chick, grabbed it and carried it to the place where the adult birds had disappeared into the bushes. This was the best he could do in our eyes, even though I usually take a "non-interventionist" stance. But the fact was that if the humans had not been in the way, the family would not have been separated at all, so the farmer here had only repaired what had almost been destroyed by the unfortunate meeting of cranes and humans.

But why did the crane parents lead their chicks into the bog birch bush forest in the first place?

Already in the evening, we could observe them all together on a freshly mown meadow 300 m behind the forest, where they were to be seen daily from that point on. They had found a new roost, a waterhole right next to the meadow, in the damp birch forest (only recognizable for the very careful observer).

If the "Pioneers" had regularly changed their feeding meadow after two or three weeks in the earlier years, this experience would not be worth mentioning. But as it was the first time, it is worth a second thought: The late breeding had created a new problem, one that the "Pioneers" had never faced before—the grass grew faster than the young, they were never completely visible, and the fox could hide all too well in the grass.

This problem was evidently made even more clear to the adults the more often the fox suddenly appeared in the vicinity of the chicks, and possibly just at the point when one of them was on an exploratory trip through the grass forest.

The thought suggests itself to me to interpret changing to a different foraging meadow as a solution to a problem after analysis of said problem has taken place. If the readers want to follow me in this, however, a new problem opens up for us: How did the cranes know there was a meadow behind the dense bog birch forest? I find this question easy to answer: Cranes (especially our experienced "Pioneers") certainly know their surroundings very well. But how did they know that the other meadow would not present the same problem, but that it had been mowed the day before?

Had one of the mates possibly flown off on special missions to investigate? I had already observed that, although the adults normally never leave their family alone, in rare cases, one of the parents would disappear for a short time, up to half an hour. It is not at all improbable that the cranes heard the tractor working, as the distance was about 300 m. Maybe the male flew over to check out what was going on. But how did the mate, perhaps noticing on his survey flight that the meadow nearby had been freshly mown, communicate to his mate that there was now a solution to the tall grass problem?

Do cranes communicate with each other at all, and if so, how and about what? Unfortunately, our observation possibilities are very limited. Moreover, our understanding of the birds' behaviour is patchy. The numerous gaps in our knowledge are held together by only a few scraps of conjecture or knowledge. One particularly foggy hole of ignorance is the question of how cranes communicate with each other.

Let's start with something apparently simple: the question of how two cranes manage to agree on a joint and simultaneous departure. First guess: "That's easy: One, maybe the dominant one, flies off first, the other just follows him." But it's not like that, at least not in most cases.

Anyone who has the chance to carefully observe crane pairs and groups taking off on a sufficient number of occasions will see: In by far most cases, the two cranes actually take off at the same time, not one after the other.

So, how does the joint take-off proceed? One of the two partners stretches its neck, extends it forward, so to speak, in flight posture and flight direction, and then trips a few steps in a very specific way, as if it is already taking off.

When the other partner also wants to take off, he or she joins in this short preparation for flight, and after a few seconds of coordination, both take off.

At first, it remains open as to how the one partner knows that the other one will join in, because, in most cases, I have observed that they actually take off together. Do they communicate beforehand through soft sounds that are inaudible because the distance is too great for us to hear? Or is the body language described sufficient? More about that in a later chapter.

Another situation that requires communication is when the two partners want to do something different. Let's take a typical case: The "Rattlers" (I called them this because of the unmistakably rough voice of the male) had a courtship early in the morning, walked around for a while, and then stood around indecisively. I wondered why they didn't start feeding. Eventually, one mate went off to the right towards the edge of the reeds, and had almost disappeared when he stopped—the other one had not come with him.

This one stood on the left about 50 m away without moving. Now, both stood and looked at each other. Finally, the one on the left turned away from its partner and very determinedly continued to the left—the second one had to follow if they didn't want to lose each other. Shortly afterwards, the left crane took off, clearly having the initiative, and the second then quickly followed, gaining a few metres in flight.

Such situations, disagreements, so to speak, are rare in my opinion. Misunderstandings are more likely to occur. As such, I observed a crane that took off, apparently in the certainty that his partner would come along. After about 100 m, he noticed that his partner was not following (had it looked back? Does the partner flying in front regularly make sure the other one is following?). He turned a loop, flew back to the starting point, and flew over his partner, who probably, as a result, felt animated to fly up and also took off.

I experienced another instance with an apparently young pair (at least, it was new to our area) that was looking at various possible territories and foraging areas. I observed it one day in the northernmost pasture of the Brooks. From the body language of one, I suspected they were about to take off, but nothing happened. This crane made several attempts to use the signals I described to tell its mate that it wanted to take off, and to "pull" it along. After five or six attempts, they both took off, but only for about 100 m, after which the second partner descended back to the meadow.

Here, the game was repeated. One showed that he wanted to fly off, the other one joined in listlessly after quite a few attempts. So, they arrived over the course of three short stages, very close to me, and I thought that they would now fly over the woods to my left. I suppose the first crane intended to do so, but the second, the listless one, settled down by the pond. He seemed to be saying, "Why don't we stay here? It's okay here!" The other realized this belatedly, turned around, and flew back to his partner.

They stayed for a while and only flew off much later. Their destination was the "Little Moor", where a one-day territorial fight took place the next day,[6] something that I will never forget: My son Bengt and I were sitting at different places in the Brook, each of us watching something interesting—he the hunt of a white-tailed eagle, myself the territorial fight of several cranes. We were communicating what we were seeing to each other by radio, when we suddenly both fell into the style of radio reportage where two commentators describe a sporting event in two different locations and take their turns (when our first soccer league has its Saturday afternoon matchday, there are several games at the same time, with a radio commentator in each stadium, and they take or get their turn). So, each of us reported to the other the exciting moments, along with accompanying side stories, as if in a "stadium", reformulating the scenes into a sportive competition. I began as follows:

... and now again, directly from the 'Little Moor Stadium', which the spectators and I myself love so much because its builders have left it close to its original natural appearance, one always has the impression of being outside in the wild during these regular competitions. But that's already where the territory defender comes out of his hiding, so we won't dwell on trivialities now, but try to convey to you, our dear listeners, the excitement from the fighting arena. A moment ago, one might have thought that the new team of two, which is competing on this turf here for the first time today, could conquer the turf without a fight, for, as we all know, for the defenders, the famous 'planner duo', this is only a secondary turf, a substitute turf, perhaps even only a second substitute turf, but now the duo is handicapped, for one of the partners is already sitting on her eggs and cannot take part in today's competition, unless, for her part, she is willing to leave their home turf to the ravens and jays, but she doesn't seem to want to do that, no other way for us to interpret the fact that the male represents the duo alone here today—surely not an easy task, especially as they seem bogged down with three territories anyway and can't be everywhere all the time—but there it goes, the 'planner' goes on the attack. So far, he's walked a measured stride around the attacking duo, who haven't even identified themselves; is this the still new 'seeker duo' or a completely unknown team? The 'planner' has been showing off all of his power and experience, neck stretched out, beak up, but now he shoos the two in front of him, the female attacker shears off (a clever move,

---

[6] Here is another territorial dispute from a completely different location, lasting about 25 min, which starts very slowly with unison calls from a distance and consists mainly of "walking around each other" until the stronger pair finally drives away one of the partners of the weaker pair (who are presumably younger, as you can hear from the voices): http://bit.ly/2qgaNgl; this is a 7-min excerpt, edited from a longer recording by Bertram Preuschhof, who pointed me towards this video.

because now the 'planner' can only chase one attacker), they fly close above the ground, one behind the other, in small and large circles, land again and run out easily, yes, there is elegance to be felt, all of the experience of the old 'planner', but also the unselfconscious youth of the attacker. The combatants seem to be allowing themselves a few minutes rest, no, here we go again, but interesting, the 'seeker' female is poking the ground. What is she looking for? Or is it a distraction, a psychological ploy, sort of taking possession of the territory? Now, all three are poking around in the grass, but only briefly. This time the 'seeker' male summons up all of his courage, stretching up his neck and his beak, the 'planner' does the same—we hold our breath. You, dear listeners on the radio sets at home, may be able to hear the tension crackle, no, it is not, as you may think, an atmospheric disturbance, but really and truly the crackling tension: They walk side by side, as if in slow motion, eyeing each other, seeming now to mark out a boundary with the path, 'to here and no farther', is the fight over yet?

No, it was far from over; it went on for a long time. The "Planners" male tried to hack the "Seekers", becoming more and more aggressive, but didn't really catch any of the competitors. The apparently established border was soon lifted again, as the territory owner didn't want to give up, wanting to drive the intruders away for good after all. Our radio transmission continued back and forth; Bengt reported from the site of the white-tailed eagle's hunt amidst the scared and startled herons and geese, I from the "natural stadium". It was exciting, spontaneous and silly, for example, when a number of other cranes flew in but didn't take part in the territorial fight, rather becoming "cheerleaders", or when Bengt called out: "What tactics did the two parties use to get out of the half-time break? For the geese, it can really only be to come out with their skins intact, whole or not at all!" The turf war in the "Little Moor" was ultimately won by the "Planner" (Fig. 4).

And there was another "discussion" among cranes that Bengt and I observed the following year from different places, so that we, connected by radio, could follow a longer flight: The crane pair, probably still young and without a territory, had already caught our attention in the morning because they were flying aimlessly around, occasionally landing and then flying on again. Suddenly, I saw only one crane from this pair fly north; eventually, the two came back from the north, having looped around and met up again. They flew a few laps over the central moorland area, then turned east, then shortly afterwards—always "discussing" through a few contact calls—set down to land. However, it appeared that only one wanted to land; the other one seemingly disagreed, climbing back up, calling out and dragging his partner on with him.

Fig. 4 Common Cranes engaged in a territorial fight; this is not the same situation as described above, but this photo was taken in late winter in Sweden and involved more serious attacks; it is presumably a fight between males, while the females watch, the female on the right calling supportively and spreading her wings demonstratively. (© Carsten Linde)

I believe that I witnessed a decision-making process here, an argument. Perhaps it was even about the question of which territory could be occupied, with one crane possibly wanting to go to the "Little Moor", the other to the pond in the northern part of the Brook. I will address the decision-making process in more detail a little further along. Before that, a few examples showing that the crane communication technique is far from perfect (just as, I am sure you will agree, our human way of communication is anything but perfect).

I believe that cranes may not have enough different and sophisticated ways of expressing themselves. I suspect that they perceive more than they are able to communicate to their mate, and think more sophisticated thoughts than they are able to communicate. While we humans are undoubtedly far ahead of cranes in thinking and communication, the phenomenon I am addressing here for cranes is at least partially familiar to us: We sometimes know exactly what we want to say (or write), but we don't succeed; our partner or fellow human misunderstands us, we can't find the right words, or our body language says something different than what comes out of our mouths.

Therefore, I can well imagine how misunderstood the crane felt in the following example: At the edge of one of our crane territories, practically at the border of the nature reserve, there was a small primitive high perch from which I had a good view onto two meadows farther back. There, the "Sly Dogs" used to search for food in the morning. One morning, I wanted to observe them more intensively, which is why I chose the hunter's blind. As always, I had my binoculars with me.

But the cranes did not behave according to my plan. Still in the twilight, they came flying exactly in my direction—at first, I was pleased, because I could observe everything with the binoculars particularly well. But then, they flew closer and closer to me, finally landing directly in front of my high seat, where I had never seen them before. Now, I was in a quandary. Instead of watching the cranes without disturbing them as I had planned, I was now only 2 or 3 meters away from them, with no more need of binoculars at all; but if they stayed here all morning, I would involuntarily take root here, or at least reap a stiff spine, if I didn't want to frighten them.

Only a little later, one of the cranes spotted me (I was visible because it was a very simple, open high perch, with only a seat plate; it had no facing of any kind, let alone a roof). The crane looked directly up at me and made a slightly excited but not particularly frightened soft "gurr" sound, something like "prrr" or "drrr". The other kept poking the ground and replied "drrr", sounding very pleased. The first one went "prrr", with a slightly warning undertone (although that may have only been my interpretation), while the other one kept poking reassuredly and replied with "drrr".

So, it went back and forth. But even the first one could not quite figure out what I was all about. Perhaps he recognized me as a human being (from whom it was actually necessary to flee), but he had never seen one so high in a tree before. If I seemed threatening to him, he probably wondered why I didn't move. It was obvious to me, though, that he was trying to direct his partner's attention towards me, something like "Look up there, what's that?" But he didn't succeed.

The crane solved the problem in a different way: After a few minutes, he had obviously realized that his partner could not be dissuaded from foraging, so he pecked around a bit in the meadow in his area, keeping an eye on me the whole time, "cooing" and clearly retreating step by step from the perch, always looking back at me. His partner followed, probably oblivious to this change of direction (and still oblivious to me). The pair walked across the meadow, apparently as usual, searching here and searching there. It seemed to me that one of the cranes had pulled his partner "purposefully" away from me.

In thinking about this observation, something struck me. When we humans fail to get our partner to look at us with words ("Look what I have here"), we nudge them. The crane didn't do that. He didn't nudge his partner.

Never—except during copulation for a few seconds—do cranes have any bodily contact (as far as I have been able to observe so far), not even when sleeping in the water, where they have just enough distance from each other so as not to touch their neighbour's feathers. They never ruffle the head feathers of their partner (which some other birds do). They usually don't really peck, but only hint at it, even though, in rare individual cases, bloody fights have been observed. So, physical contact is never a means of communication in cranes, neither in social or mate bonding nor in fights. Body language, however, is.

A second example of insufficiently differentiated communication possibilities was a recent observation of a crane family foraging directly along the path on which we (my partner and I) were walking, only 10 m away from us: the male (the somewhat larger crane), having noticed us, strode swiftly towards the nearby bog birch forest, constantly looking around at his partner and his juvenile; the female hesitated, but also moved away from the path, while the teenager carelessly—as children do!—continued looking for food or playing on the ground only a few meters away from the path. The male did not succeed in luring the other two away from us towards him; the female succeeded only gradually in luring the young bird away from the path, while we moved very slowly, mute with excitement at such a close encounter and anxious not to frighten the cranes (Fig. 5).

Larger groups of cranes communicate with each other, for example, when flying. When a flying crane group wants to land, you can always hear excited calls. But what are they "discussing" in these instances? One day, I observed a lengthy "discussion": On a day still early in the season, a group of perhaps 35 cranes was coming from the southwest. They flew over the area of the "Sly Dogs" and had clearly changed their flight direction to the east.

Since it was already late afternoon, almost evening, I wondered where this large group could still fly to find a suitable place to roost. There were possibilities in the Brook that were never used by such large groups in the spring,[7] but perhaps, in view of the advanced time of day, they served as a better solution than to fly to the Schaalsee. So, I wondered about the thoughts of the cranes and about which place they perhaps had "in sight" as a roosting place.

I followed the group with binoculars, and even before they had disappeared over the trees far to the east, there was suddenly confusion in the flying group. They stopped "in mid-air", as it were, flew in small circles around each other

---

[7] At least, I didn't know of any having done so at the time. In 2018, however, a group of about 24 individuals, both juveniles and adult non-breeding cranes, stayed in the brook and probably occasionally spent the night in the central moor pond, residing for most of the rest of the season in the mill pond a few kilometres north of the brook.

Fig. 5 Several years later, I met a juvenile crane who was only little more than 10 meters away from the path in the woods. This crane was probably 4 months old, already fledged, I had seen one of its parent birds a minute earlier deeper in the woods, 20, 30 meters away from me, but the youngster was experiencing its own way, so we met: He waited and looked at me, I waited and looked at him and very very slowly raised my camera. (© Bernhard Wessling)

and then turned their flight direction by 180°. Ah, I thought, they were worried about the same thing as I had thought, namely: "Will we actually make it to ... (to the Schaalsee) before dark this evening?"

Now, they wanted to land in the area in front of me, which, however, belonged to the "Sly Dogs". These protested promptly with active unison calls, so that the landing attempts were aborted and the group regained some altitude. I heard the contact calls or flight calls that the birds exchanged among themselves. They continued to circle indecisively directly above me, calling all sorts of "words" to each other, flying a little to the west, then to the east, and finally making a decision and heading off to the north. There lies the Nienwohlder Moor, and I suspect that they spent the night either there or, more likely still, in the Mill Pond in a small spot called "Graeberkate" not far from the moor near Bargfeld-Stegen. There is plenty of space in this small lake, with its flat end where one can very often find cranes in late evening and at night.

As limited as the cranes' communication skills may be, they are obviously sufficient for survival, as well as for coping with difficult situations. I once observed a crane pair (long before I got the impulse to observe these various occurrences) that I called the "Movie Stars", because TV footage was once shot there (although it may have been the pair I otherwise called the

"Planners"). I knew they liked to come out of the edge of the forest into the clearing in the afternoons, where they foraged and took their already quite large juveniles out.

I had promised my then-young children that we would "definitely" see the cranes, and even their offspring, for the latter were now big enough. I stood with my sons in a suitable place, but they were disappointed, for the crane family was not there; they would so much have liked to have watched them through the telescope that I had set up.

After some waiting—nothing for an eight- and a ten-year-old!—the cranes finally emerged from the forest, but without their juveniles. My sons asked, "Where are the little cranes?" But I didn't know either, fearing the worst. After a few seconds, the adults stopped, turned and looked back into the forest.

Nothing happened. Finally, one of the cranes went back into the forest and came out with the two youngsters a minute later. I remember explaining to my sons: "That's just like you two, look, they were probably up to some non-sense in the forest when they were supposed to come to the meadow to forage but didn't follow; so the father had to go and get them," but I wasn't at all aware of the significance of this event at that time.

It wasn't until I was writing this chapter that this earlier observation came back to me, and I thought about its relevance. Several mental feats have taken place here: First, the adult birds noticed that the young were missing (did both adults realize it at the same time? or first one of the two, who then told the other, for example, by standing still, "Something is wrong here"?). Anyway, being the parents, they looked after the well-being of their kids, guided and fed and protected them, and thus did not want to leave them alone (Fig. 6).

Then, they realized that their juveniles had not come along, likely engaged in a very minimal amount of communication ("Stop, something is not ok"), waited at the edge of the forest and looked back into its interior. Now, they assessed the situation: If their offspring came out on their own, everything was okay, and thus they could continue together as a foursome. But if they did not come out, something would have to be done. I don't know whether the young birds were visible to the adults, whether they could observe what the little ones were doing. Or whether the young ones, as always, not worried about losing their parents, were exploring everything; worrying, even in cranes, is a one-way road that only leads in the direction from parents to children.

After assessing the situation, perhaps the parents discussed among them-selves: Who goes to get the lost family members? One went off, probably the male (being a father myself, I can imagine this from my own experience).

In the forest, he had to somehow persuade his youngsters to come along. He managed this with the help of methods of communication unknown to

Fig. 6 A Crane pair just emerging from the woods with their chicks, still very seclusive. (© Bernhard Wessling)

me, the kids followed him, and the family went on to their afternoon foraging. My sons, for one, were happy that the young cranes weren't dead. I'll admit it: I felt exactly the same way.

Another observation concerning communication: Near the nest of the "Foxes" (the origin of this name will become clear later as well), a white-tailed eagle was sitting in a tree. Perhaps this was new to the breeding crane, perhaps this hunter's lookout was just too close and felt threatening—in any case, I suddenly heard short, restrained calls from the nesting site. I thought maybe the crane sitting on the nest was trying to call the mate (who I knew was standing in a meadow—not visible from the nest—about 800 m away). The calls were repeated, becoming louder, more insistent. They were single sounds, like "pööt", sounds that I had not heard before.

It took at least 5 minutes, but the partner finally appeared and flew to the nest, after which the calls fell silent, even though the eagle remained perched. The partner sitting on the nest was obviously reassured.

Another incident, unfortunately, did not have such a happy ending: In the year in question, as in the 2 years before, the "Pioneers" had lost their brood, and they had started another breeding attempt. One evening in early May 1998, I visited the crane guards. As usual, I saw a single crane in a certain meadow on my way into the Brook and said to myself, "Good, the 'Pioneers' are still breeding. Five more days and the chicks will hatch."

I met the crane guard and we discussed the status, which, unfortunately, was not good: Only one pair had hatched one chick; all others (except one) had already stopped breeding at least once. The "Foxes" had begun a new attempt. The "Rattlers" were perhaps still breeding, or not. The "Sly Dogs" and the "Seekers" had also lost their broods. At least the chicks of the "Pioneers" were about to hatch. While I was walking back to my car, I heard a short, not very cheerful unison call from behind a copse, pretty much where I had seen the single "Pioneer" standing earlier. Had he been driven away by another pair? By which one, then? What pair would be calling here?

First, I saw nothing. The birds had probably flown away again. After a few minutes, I spotted them: A pair was walking resolutely northwards on a pasture where several horses were standing. This was a meadow neighbouring the one the "Pioneers" usually used. One of the cranes walked about 50 meters in front of the other, and even flew a bit, seeming to want to take off. Both walked between the horses as if they didn't care (cranes would never normally walk between horses or cows). Had the "Pioneers" just stopped breeding? But one of them had been visible at the edge of the forest just then, so the other one must have been sitting on the nest.

Only in retrospect did the complexity of this situation become clear to me (if I am interpreting the events correctly): One partner was foraging while the other was still sitting on the nest. Whatever had happened at the nest, the breeding partner must have realized that the brood was not going to live and flew to his partner, whom it could find safely and quickly thanks to the limited choice of meadows.

It arrived there, and there was a unison call indicating that the male partner apparently understood at once what the situation was; they were aroused and began marching right through the middle of the herd of horses, with no destination except away. This was extremely unusual: Never before or since have I observed that cranes would walk in between a herd of horses or cows; they will always keep a safe distance. Why then this time? My interpretation is: The brood was finished, the partner who was not the last one on the nest was "upset" or at least disappointed, to the extent that it did not care about the horses any more.

The crane guard with whom I had spoken a few minutes earlier found out the following day that it was indeed the "Pioneers", and that they had given up on the brood again. Since we didn't know the exact location of the nest, and since there was another crane pair breeding fairly close by, I decided not to look at the nest and hence not to investigate what the reason for the brood loss was.

But how did the partner who was still sitting on the nest at first inform its spouse that the brood was finished? Or was the simple fact that it had abandoned the nest sufficient to convey this complex information? But how does the absent partner then distinguish this communication from what was observed and related to me by a biologist from Osnabrueck? He spent a lot of time in a hide very near a crane's nest for his graduate thesis. The crane sitting on the nest showed clear restlessness after many hours spent there alone. It called its mate (much in the same way as I had heard the one sitting on her nest below the white-tailed eagle), but was not heard. Finally, it flew off, leaving the clutch alone, to the horror of the observer (because it could easily be that some ravens would detect the eggs and take advantage of the content). Shortly afterwards, two cranes returned, one sat down on the clutch and the other flew out—presumably to finally find something to eat as well.

How did this breeding crane communicate to the still foraging partner that it was time to change places and let her look for food while he took over the duty of sitting there? How, in the event I described earlier, did the breeding crane communicate with her partner that the brood was lost? And isn't it surprising, shouldn't it make us think and research more deeply when observing that another crane sitting on her nest, feeling scared by the white-tailed eagle above her, was able to call its partner from a 800 m distance to "come back and help", and the partner actually came!?

Many questions still remain unanswered. Let's take a look at the life of a young crane. In the first year, it moves with its parents to Spain (or southwest France or North Africa or even Ethiopia) in winter. Either while already there or, at the latest, after the return to the home climes, it will separate from its parents, voluntarily or by force, because they reject it. Often, one can observe this process here in Germany at the end of February or the beginning of March: The adults come back to the breeding territory and have a third crane (or two) with them. A few days or weeks later, they drive it away—it still wants to return, seeks closeness, but is consistently rebuffed like a stranger; it is now on its own.

By the end of March, all of the experienced pairs have normally taken over their territories and are already defending them. Then, relatively large troops of five to fifteen cranes suddenly appear; in 2019, I observed as many as 65 individuals in the Brook area, and in 2020, up to around 100. These are young cranes, often still single, and often grouped together with non-breeders, and adult cranes that either didn't find a territory or have voluntarily taken a break from breeding. They hang around, stir up a lot of "alarm" in the areas of all the other cranes, and make clear that they want to spend the night in the most beautiful places, but the respective territorial pairs do not allow this (if they are strong enough). How did these young cranes come together?

Later, these groups break up. One sees groups of three, four and five. Soon after, more and more cranes are flying around in pairs, or you can see a group clearly consisting of a number of pairs. Are they young couples? Lovers? Engaged? How do they meet? How do they decide upon each other?

The former Brandenburg Minister of the Environment, Eberhard Henne, was also, for many years, the head of the Schorfheide-Chorin Biosphere Reserve. From his experience with banded cranes, he can now say that some cranes that were ringed when they were not yet fledged looked for a new territory just a few kilometres away from their home territory. But what even he does not know is: Who were their partners, where did they come from?

Among our pairs in the Brook, I noticed that individual young pairs were actively investigating territories.

Years ago, there was a pair that appeared for the second time in a meadow that had not been visited by cranes for years. Although it was under our constant observation, we simply never saw any crane appear there.

This meadow, which I'll call "Light Pasture" here, is nice and secluded, quiet, probably also rich in nutrition. Maybe red deer and fallow deer disturb the cranes sometimes, but there are busier deer spots where the cranes can nonetheless hang out, forage, and still avoid the deer.

In 1997—in May, after the time when the groups of juvenile cranes had dissolved—a pair appeared on the "Light Pasture". True, this was too seldom an occurrence for us to speak of a territorial occupation yet. But if my hunch was right, we would be able to observe a new pair there the following year.

In fact, a year later, a new pair, the "Seekers", found their adopted home in this territory. One of our crane watchers was able to observe an extensive courtship, unison calls, copulation and the subsequent dance. I had already noticed the first signs of a territorial occupation at the end of February; this was the confirmation.

Sure, we didn't know exactly if the "Seekers" were really the pair that we had already seen in May of the previous year on the "Light Pasture". But it was to be regarded as probable, because, after all, we had had no pair here at all for years when that pattern was broken by the first territorial investigations in May of the previous year and then the courtship and copulation in the following year. Still, individual recognition of the cranes was not possible for me at that time.

The question I first asked myself was: What criteria does a young crane pair use to make their territorial decision? They met, decided upon each other as mates, however that happened we don't know (we usually don't know how it happens for ourselves, much less understand how someone else, say one of our family members, could decide upon this man or that woman, this boyfriend or that girlfriend; the last one was so much nicer and a much better match, weren't they?).

Then, the pair looks at one potential territory after another. With a little luck and a little more time, it is possible to observe a situation like this. You see a pair in a territory a few days in a row, they walk around, they fly in and out, they look for food at the edge of the territory. After a few days, they disappear, but then a crane pair reappears in a completely different place for a few days—the sequence of these events suggests that it is this first pair that is searching for a territory and looking at potential breeding sites.

Now, it is May or June, and the birds have chosen a territory—perhaps according to criteria that we humans would find just as important for crane territories: a sufficiently moist nesting area, good opportunities for retreat, a meadow in which to guide and feed the chicks, distance from hiking paths and roads, etc. So, is there a consideration process going on in the birds? Are both of them considering the Pros and Cons? And how does the joint, assuming it is joint, decision-making process proceed?

Perhaps it is simply by virtue of the fact that they both "coo" contentedly, that neither of them wants to leave or does leave, that they notice from the other's behavior that the other one is just as comfortable coming back to this place in the evening.

Territory selection cannot be an inherited trait, because, after all, it happens differently in each pair and each territory. It must be an individual decision-making process that takes place over months and years, and sometimes in areas far away from each other. So, the cranes must be somehow "aware" of themselves and their mates in their environment, perhaps even of their future as a breeding crane pair. Certainly not "conscious" in the sense of "self-reflection", but still to the extent that decisions like "yes" and "no" or "maybe" become possible on the basis of observations and judgments. What mental processes are going on here?

Obviously, animals have criteria by which they judge the suitability of a territory. And they are able to remember their observations, to "think about it" by comparing. It is only through comparison that judgment and subsequent certainty about the appropriateness of the decision made comes about. Based on good or bad experiences, decisions can be revised; we can clearly observe this as well. So, the cranes have a "concept", a picture of what a territory must look like—a picture that develops through experience.

Now comes the next winter. Do the two young cranes know that they will be able to breed the following year or the year after, that they will be sexually mature then? They fly south, they come back—how do they agree to fly here again, how do they find their way?

One crane pair that I observed for years (and of which I could previously, until I had a way of really knowing, only assume that it was the same pair every year) inhabited the most densely overgrown, partly very wet, trackless, and therefore most difficult to access, area in the Brook (which may only be entered with a special permit such as I had as a voluntary nature conservation officer). I refer to it here as the "Green Brook".

Opportunities for outside observation were very inadequate, providing us with little information. In one year, I was sure that the pair had again failed to fledge any young, if they had bred successfully at all. The reason for this, I suspected, was that the terrain between the breeding site and two possible foraging meadows was simply too treacherous, particularly as it also included a fox den. In front of this burrow, once when I was deep in these wet woods, I was able to observe three fox cubs playing in the sun. It was such a wonderful experience, although, of course, it was clear to me that these fox cubs would soon learn, among other things, to prey on and eat crane chicks.

I cautiously approached the center of the cranes' territory. And this time, I managed to spot the cranes before they saw me. They were again in their barely sufficient foraging area (and, as expected, without juveniles). Now, I watched as they slowly moved east (I was coming from the west). They crossed the border into the neighbouring territory, which had become available due to a dramatic incident (which I will relate later), and slowly and carefully conquered the most important and best feeding area there.

I had actually expected this earlier, and this is what I had hoped to observe. I don't think I witnessed the first act of conquest that day; there had probably been such attempts before. I suppose the pair were gradually testing the area to see whether they would be driven out or not.

Subsequently, I only observed the pair in the new territory. It took over the roosting place abandoned by the previous owners. This area is one of the most beautiful wetlands in the whole Brook and had been used as a breeding site by the now-absent pair in earlier years. It was soon clear that the new occupants would not be taking over the previous owners' last nesting site. While the pair foraged nearby, it never went in the immediate vicinity of the orphaned nest site (at least not as far as I know). During the time we suspected a nest would be built, it often flew to the roost during the day. There was what I considered to be a very suitable nest site there. Certainly, the pair had already begun building the nest, for at times, it was not to be seen for a long time. It was probably working on the nest somewhere near its roost.

I really wanted to know where the nest was so that I could draw it to the attention of the forester and the farmer, who regularly had to work there, and ask them to be considerate. From a distance, however, I could not spot the nest site. In no case did I want to tramp through the area—that would have been an irresponsible disturbance from my point of view.

One day, I noticed a crane fly out of the wetland area of the roost and turn north in a large arc to look for food. On another occasion, I watched it enter the area from the north, describe an arc, fly behind and along a tall earthen wall hedge so that I could barely see it, and descend into the area of the roost.

Shortly afterwards, I saw another crane leave the wetland. It too flew along the earthen wall hedge, approached the forest line at a low altitude, elegantly gained height, flew on just above the treetops and disappeared from my view. Ah, it was brood change: now the other partner was going to eat, and in a different location at that. We have seen this in many crane pairs: For example, one partner flies north to a certain meadow, the other perhaps prefers to fly east.

A few days later, my own crane guard week began, and my son Bengt and I spent the time together again. Among other things, I had planned to keep a closer eye on this particular pair and find out exactly where they emerged from the wetland where their nest was—because that would later be the place where they would lead the chick out. If it was an agricultural meadow, I would have to inform the farmer and make arrangements with him.

I did not manage to determine at which location the cranes left the wetland. Again, I saw one of the two birds only when it flew behind and along the earthen wall hedge. This was strange, because this time, I had positioned myself on the other side of the row of bushes. Had the crane chosen a different "exit" because it had been watching me and wanted to avoid my gaze? And did it generally like to fly along behind the hedge so as to be out of sight? Did it know that I would be able to watch it if it took its usual path along the bush line? And was that why it flew behind the hedge line, because it "knew" that I would be able to observe it if it flew in front of the densely overgrown earthen wall, but that it was difficult or impossible for me to observe its flight if it was behind it from my standpoint? Maybe it was coincidence, I couldn't tell.

I resolved to be at the same place very early the next morning, setting out before dew and day. I was in place early enough—I thought. But instead of hearing the morning calls of the cranes just before dawn (which were just sounding in the other territories, as I could hear from a distance and through my son's radio), a little later, I discovered one of "my" two cranes coming from the north on its return flight from breakfast! Man, he was up really early!

Now, I hoped he would make his bow and fly in along the earthen wall hedge so that I could see (because he couldn't possibly know I was here; I had come while he was having breakfast outside the area). I assumed that, this time, I would succeed in determining where he would land and then walk to his nesting location. Sure enough, he flew his bow but stayed aloft, not descending as usual.

He was heading towards me. I watched him through my binoculars and could see him tilt his head slightly to the side and look down at me—clearly, he had spotted me from a distance, although I had thought I was well hidden! "Bad luck for me that he just has very good sight, better than mine," I thought. Now, he flew a wide loop above me and looked down at me every now and then. Then, he flew back to the hedge line, descended and landed somewhere in the wetland, where I could not see him. A few minutes later, the standard call sometimes heard during a brood exchange rang out—but strangely: It seemed to me that it was not the typical unison call, but rather a single call in the melody of the duet, a "solo duet" so to speak, the duet part of the male performed by him alone.

I could not explain this observation. My first guess was that only the crane that was sitting on the nest was calling, while the other one that had seen me remained quiet, saying "Shhh!" to its mate, so to speak.

Some time later, through further observations, we came to suspect that a breeding site was located about 400 m west of the wetland used by "our" pair. It was hard to imagine that another crane pair was breeding there—we should have seen and heard these birds. It had to be "our" pair, which was not breeding on the eastern edge of its territory, but 400 m farther on the western edge.

This allowed for only one conclusion: They had deceived me. Visible from afar, they flew in in a wide arc to the east and stealthily crept west to the breeding place. We received confirmation of this hypothesis when the pair with their chick emerged from the thicket for the first time after hatching: to the west of the territory, 400 m from the site initially suspected. Another 2 months later, during a follow-up search, we found the nest. The pair had employed a whole series of deceptive maneuvers: Low flying, "walking" and hiding. But most impressive was the mock welcome call as a unison call, which was only half a duet, a "solo duet," just the male's voice singing the duet's tune.

For 3 weeks, this crane pair had succeeded in deceiving me "very slyly" about where their nest was. Only through very close observation had it been possible for me, together with other crane watchers, to determine the correct location of the nest. From now on, I called this pair the " Sly Dogs".

The episode gripped me: How purposefully had the cranes planned and executed these deceptions? Above all: What intelligence and consciousness was behind the "half" unison call that was possibly deliberately set up to deceive me? After all, not all cranes try to hide their nest from humans with so much effort, but only those that live near humans, i.e., those that are observed or feel observed (cranes primarily hide their nest from other predatory animals, such as foxes and wild boar). Can a creature that is not aware of itself actually actively hide from another creature? How does a crane know that, when it flies behind a row of bushes, it cannot be seen from the other side? How did "my" sly crane know that I heard his half of the unison call? And that, if I heard it, I would conclude that that was where its nest was?

On the subject of "hiding", a comparison with the behaviour of small children is also interesting. In early childhood, they believe that if they hold their hands in front of their eyes or pull a towel over their head, the parents can no longer see them, because they themselves can no longer see the parents. They cannot yet imagine that the parents see anything other than themselves. Only later, when playing hide-and-seek, do they make sure that they themselves really cannot be seen, while sometimes (e.g., through a slit in the door), they observe how they are being searched for.

In the following year, the "Sly Dogs" chose as their nesting site the very spot that I had mistakenly thought was their nesting site the previous year. Ah, I thought, last year they checked whether the water level was stable during their numerous stays here. (I believe that experienced crane pairs only nest in places where they know or suspect the water level will be stable for long enough—it's part of their life experience). So, maybe they hadn't "intentionally" deceived me at all. Instead, perhaps what I had observed several times was simply one or both of the partners looking at the potential nesting site, especially since it was a roosting site. Even if this were true, I still found it interesting, because it suggested even longer-term planning. Planning more than a year in advance, choosing alternative sites!

In summer, however, I observed that the path from the nesting site to the feeding meadow was all too arduous for the chicks. So, I discussed with the forester that this now overgrown part of an unused meadow should be mown once, and he agreed. At the same time, we had agreed and coordinated with the responsible authorities that we (some nature conservancy group members and I) would fill in a ditch, which, according to the Schleswig-Holstein nature reserve development plan, should have been dammed up long ago. This created a new wetland area where I hoped another crane pair might settle. It formed more quickly and grew larger than I had expected.

Already at the end of February of the following year, when I had otherwise not observed any cranes at all, I saw a crane pair right next to the new wetland. Was it the hoped-for new breeding pair, the "Seekers", or had the "Sly Dogs" expanded their territory?

I asked the other crane guards to look very closely, and also tried myself, through intensive observation, to recognise the behaviour patterns of the cranes in the new wetland and compare them with those of the pairs I knew. As time went by, it became clear that there was still only one pair in the extended wetland. It was breeding in the new part, occasionally came to the old breeding site "East" and had changed its feeding habits.

Unfortunately, even that year, it was not possible for me to determine without a doubt which crane pair it was. If it was the sly territorial pair, they were really clever birds: They had flexibly changed their behavioural patterns, adopted a new wetland and abandoned the breeding site "East" (because of scrub encroachment?).

From the cranes' point of view the new wetland was an excellent breeding place: If someone approached, the cranes could quickly disappear and make themselves invisible—there was no better place in the whole Brook for this. Unfortunately, after the breeding season in the autumn of 1998, the Hamburg forester, concerned about his deer and their hooves, tore down our dams (which lay exactly on the border of his territory with Schleswig-Holstein), thus draining the new wetland again. The consequences could be seen the following year: The "Sly Dogs" retreated back into the more confusing interior of their territory and drove away all potential neighbors. Everyone involved in conservation knows setbacks like this, the destruction of already successful conservation efforts through thoughtlessness or bureaucracy. I have experienced it many times and still can't get used to it.

However, this does not change the fascinating flexibility in the behaviour of the "Sly Dogs". This pair attempted a re-breeding in another year despite a brood loss in the same area of their territory, during which we did not manage to see the chicks even once while they were being led. It was not until early September, at a time when I had already ceased to believe that a re-breeding would be successful, that one of our friends observed four cranes—two adults and two juveniles—fly into the foraging meadow far back. Because we had only one other pair with only one fledged youngling in that year, it could only be the "Sly Dogs"—the masters of hiding, camouflage and deception.

Arrival in the Brook After Returning from Wintering Grounds: Alone or in Groups?

Every spring, it was exciting in the Brook: When will the cranes return? Who would hear or see them first? All of us who were active in crane protection often went to look, wait, indulge our curiosity.

Between February 5 and 25, the phones were ringing off the hook: "The first cranes are here!" Someone—the ranger, one of my fellow conservationists, myself, or a Brook visitor—had seen them "first." But one always saw or heard them only after they were already in their territory. The late Carl-Albrecht von Treuenfels, the crane-enthusiastic former president of the WWF, described in his well-known book *Kraniche, Vögel des Glücks*[1] (*Cranes, Birds of Happiness*) all stations of the migration from the north to Spain. He mentions how he himself is woken up in the morning by the crane pair breeding near his home—but one reads nothing about the arrival of a pair from the south in spring, or perhaps even "his" pair. In Hartwig Prange's book *Der Graue Kranich*[2] (*The Grey Crane*), I learned the following: "The pairs of the respective wider surroundings often arrive with the same migratory wave and often within the same migratory group. ... However, such arrival observations rarely succeed. Generally, the cranes are 'suddenly there'."

How, then, can one know in which groups they arrive? Why did von Treuenfels, who observed his cranes so intensively and identified a territory very near to his home, never describe the arrival itself? Well, the answer is simple: It takes—depending on what we define as an "approach"—only a few minutes, and the actual arrival in the territory only seconds. One is hardly

---

[1] Carl-Albrecht von Treuenfels: Kraniche, Vögel des Glücks. Hamburg: Rasch und Röhring 1998, first edition 1989 (he passed away in 2021).

[2] Hartwig Prange: Der Graue Kranich: *Grus grus*. Lutherstadt Wittenberg: Ziemsen1989, p. 58 (translation by the author).

likely to have the luck of being at the place of the event in these exact seconds. And if you are, you still don't know for sure whether the cranes are really arriving at that very moment, or whether they are merely returning from a foraging flight and had already gotten there the day before without you knowing it. So, it takes a series of fortunate circumstances and, of course, a lot of time for intensive observation to be able to observe a return from the wintering grounds with certainty.

Again and again, I thought about how the arrival of "our" cranes might proceed and whether I would ever be lucky enough to witness it. But I did not actually believe that this could happen.

Many years ago, my family and I were once again standing at the edge of a forest that offered a good view of a crane territory in a southerly direction. The pair that had their territory there was already present, as was another one. It was unclear to us, however, whether all the pairs had already arrived, because we had not yet spotted all of them—which didn't have to mean anything, because they could have been looking for food outside of the nature reserve at the time of observation, for example, or been hiding in the obscure parts of the brook.

For a long time, I had been wondering how the cranes could have found their way to the Brook. What caused the first cranes breeding here to look for a new territory so far away from both their former home (probably Mecklenburg) and the main migration route? Certainly, one could begin by saying, for the sake of simplicity: They have found no other suitable areas. But for me, this answer is too quick. How does such a search process take place?

More importantly, if migration and breeding behaviour are—as is still generally thought—hereditary, how could a crane pair from Mecklenburg start breeding in Hamburg? How did they find their way back to the migration route, which, at that time, definitely did not lead via Hamburg, but was rather more than 100 km farther east? Where did "our" first cranes meet up in autumn with their fellow species, with whom they then flew together to their wintering quarters? After all, at that time, in 1981, this first pair lived alone in the breeding area and was the furthest to the west in all of Europe. What about a little later, in the early 1990s, when there were about six breeding pairs in the Brook, still in the furthest western area? Did they fly together to the wintering grounds? Did they return together?

I did not seriously expect ever to get an answer to even a portion of my questions, certainly not on that Saturday morning when I was standing with my family at the edge of the forest. Suddenly, we saw a flock of birds coming towards us from a southwesterly direction at quite high an altitude, still relatively far away from the Brook. With the spotting scope, I was quickly able to make out: They were cranes, about 40 of them. And the "quite high altitude" told me that these were migrating cranes—the first time I was able to see a migrating flock in our area, so far to the West!

This was extraordinary, because the migratory routes of the cranes to Mecklenburg-Vorpommern or further to Sweden and Finland did not lead via Hamburg at that time, but rather (along with more migration pathways even farther to the East) exclusively via the Alsace—Goettingen—central Mecklenburg-Vorpommern line (here, the spring migration route bifurcated). This is still the case today, despite some notable changes in migration routes.

The evening before, on a business trip using the Bremen-Hamburg motorway, I had seen a group of cranes of about the same size, which had just swung across the motorway from the south into the Weser river lowlands and apparently wanted to spend the night there.

So, I said to my family that maybe what we just saw was that group of cranes from the previous evening. Bengt, critical as always: "How do you know that?"—"I mean just as a hypothesis, because it's about the same flock size. Let's see where they fly." The group flew so high that it couldn't possibly be a local movement (especially since we didn't have that many cranes in the area); it was clearly a migrating group. Odd.

The birds kept heading for the Brook in northeastern direction, but swung east less than a kilometer south of us. "It's a pity they don't fly over us," we said regretfully, because, despite the significant height, we could have observed them very well with our binoculars.

But now, something very strange happened: Two cranes detached themselves from the flock, dropped down diagonally with outstretched legs and wings, landed after a few seconds (which we could no longer observe directly) and trumpeted, while the flock, without losing altitude, moved on, only now to the east, no longer in a northeasterly direction. The landing clearly took place in the "Green Brook", a territory whose "owners" we had not yet seen this year. The group had changed its direction of flight at the exact moment that the two cranes left it.

We were quiet. This was a huge surprise. For the first time, we had observed the arrival of a pair. What's more, this pair had returned separately from the other local crane pairs, but—and this was the real surprise—not alone or in a small group from the east, but in a relatively large group from the southwest.

When I had previously thought about the return journey of "our" cranes from their wintering grounds, then it was always in the form of two alternative hypotheses: Either they came as a "brook group," as part of an even larger group that perhaps separated at Schaalsee, about 70 km as the crow flies east of Hamburg. Or they travelled as individual pairs in a larger group that arrived somewhere far east of Hamburg, from where the individual "Brook pairs" then made their way to Hamburg—just as they had probably discovered the breeding area when they first came to Hamburg in the Brook. Anyway, both of my hypotheses involved them coming from the East.

Now, I was confronted with an "unthinkable" third variant: A single Brook pair flies in a larger flock off the main migration route, offset about 150 km to the west, from the south (from Spain, via France, North Rhine-Westphalia, the Muensterland, Bremen) in a northeasterly direction to Hamburg as far as the Brook, where the group then swings to the east, crosses exactly one particular territory, and the territorial pair drops, so to speak, from cruising altitude into its territory and lands.

Was this an isolated incident? And if so—what did this event mean? Why did this group fly along a detour with our territorial pair; why did the other cranes accompany this pair at flight level until they were above their territory and then continue their flight eastwards without pause, but with a change of direction? How did our territorial cranes know that they could travel in this group, that it would fly over the Brook? Or, how did our cranes "persuade" the group to make the detour over the Brook? In other words: How did they communicate their travel intentions and the route? Why didn't they all take the main migration route, for example, to somewhere around Schwerin, whereupon our cranes could then fly west to their territory? What planning, communication and navigation skills are behind the behaviour that we have observed?

A year later, I was allowed to experience another arrival, this time I was alone. At that time, I was only a few hundred meters from the landing site, in the eastern part of the large Brook area outside of Hamburg in the Hansdorf Brook. Another stroke of luck, because it was again a Saturday morning—on weekdays, like all working people, I would have had no time for crane-watching during the day.

On this Saturday in mid-February 1996, I had resolved to find out whether the pair I later called "Romeo and Juliet" had already returned from their wintering quarters, for they had been back early the year before. The previous year, they had nested in the middle of the agricultural part of the reserve, close to a farm track, and had hatched their chicks in spite of much disturbance, but for reasons unknown to me, they did not fledge, but simply disappeared after a few weeks.

I took up a position from which I could overlook practically the entire territory, the "Alder Swamp". With my spotting scope, it would even be possible to watch the nest building without disturbing them, in case the pair were to breed again in the same place. I wanted to learn more about them this year, for example, how long before egg-laying the nest-building activities began.

But they were not there yet. Since the weather was friendly, I decided to wait. Suddenly, I saw a group of maybe 25 cranes on the horizon at a relatively high altitude, again coming from a southwesterly direction. Clearly a group on spring migration to the north, another new observation. Also, this time, the group made a turn to the east over the Duvenstedt Brook, practically exactly in my direction, but no crane pair separated from the group during the change in the flight direction. The group flew farther east and approached the area at the border of which I was standing.

Shortly before reaching the "Alder Swamp", from high altitude, two cranes actually dropped out of the group right before my eyes, with wings bent forward and legs outstretched, and were on the ground within a few seconds. I was just as stunned as the year before. It was the same story: A territorial pair were participants in a "group tour", the flock came from a southwesterly direction, which was a detour via Hamburg, changed their flight direction over the Brook to the east, and then the local pair left them and landed, while the rest of the flock continued to fly east without changing altitude.

But now, something happened that I would have been even less likely to predict. The two cranes that had just landed by no means started to feed, which could have been expected after the long journey (even if they had only flown the last leg from Bremen or from the roosting place at Diepholz Moor). Nor did they just stand there and recover from the long flight. No: They trumpeted in unison and danced!

They leapt high, flapped their wings, bowed to each other, leapt high again with wide-open wings, stepped around each other, bowed again and again, punctuated by small and large leaps, called alone and trumpeted in duet. They flew a few dozen meters here and there at a height of only a meter or two, danced again, called. They flew a circuit of the territory at treetop height, landed at the roost, flew up again. They flew to last year's breeding site, inspected it briefly and superficially, pecked around in the ground a little, danced again, flew here again, flew there, to all corners of the territory, 300, 400 meters around.

It was like a ballet, but one with tremendous music (albeit only audible in my head), nothing contemplative, but rather a succession of explosions of dynamism and power, coupled with an indescribable grace. Crane song and dance, a unique choreography for two dancers, aviators and trumpeters. And for only one spectator, with free admission.

The program lasted over half an hour. After that, the two birds calmed down a bit and ate a little, and only then took a closer look at the individual places within their territory, "Alder Swamp". It took another hour until they had really calmed down and began to eat intensively.

I walked back to my car knowing that I had witnessed something very special: I must have been the only one lucky enough to witness the arrival of a crane pair in their breeding territory after returning from the wintering grounds. Never before or since have I heard or read of a similar observation. More than that, I had witnessed an outburst of emotion, an outburst of "joy" (if I may use that human category to describe it) at being "home" again. Here was a crane pair bubbling over with the joy of life. They were happy about each other, they were happy about their territory, they were happy about the coming breeding season—this is the impression they gave me very clearly, the cranes "Romeo and Juliet". And I still feel that way today. Even as I write these lines, I have the images in front of my eyes, the powerful non-audible music in my "mind's ears", and I am still incredibly grateful for having been allowed to be there during those one and a half hours.

Since this event, I can no longer see cranes as emotionless "breeding machines", driven only by their genes to lay two eggs year after year, hatch them and automatically get the soulless fuzz that crawls out of them fledged. Instead, I recognize feelings in these animals that, apart from humans, would, at best, be granted (in an appropriate form) to dogs, horses, elephants, whales, dolphins and primates. But not to birds.

Twice now, at intervals of a year, I had seen a pair return from their wintering grounds, and a third one was to join them later. They had not flown into the Brook alone from the east, where the main migration route lies; they had not come there in a group with the other breeding pairs. They had not followed the main migration route with a larger group, arrived in Mecklenburg, and then flown to the Brook from the east, as one would expect of animals that might be capable of learning, but were predominantly controlled by their genes.

I assume by now that each pair comes to the Brook alone. Whether they are all accompanied by a larger group, I do not know. But based on the observation that three pairs each came to the Brook in a larger group from the southwest and landed, while the rest of the group changed direction and flew east, I would like to assume that these were not three unique individual cases.

The question still occupies me: How do the Brook cranes manage to find a flock that flies via Bremen to Hamburg, turns right over the Brook off to the east and flies on? How do the Brook pairs know that this group will take this route? For a better understanding, compare the migration routes on Map 1.

I should add that the Brook cranes that I saw soaring down from the groups were not flying at the head of the group, but somewhere in the middle of the pack each time. Other cranes were in the lead. What do these know about the Brook and the intentions of the two that want to land there? Or how do "our"

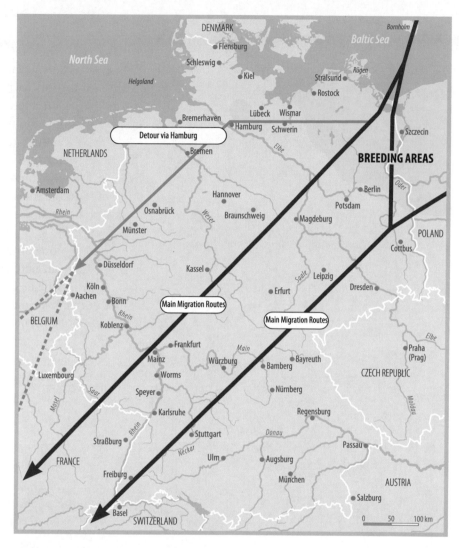

Map 1 This map shows the migration pathways in the early 1990s: main migration trajectories in red, "detour" via Hamburg in green

cranes manage to "persuade" a group of 25 or 40 colleagues that the detour via Diepholz/Bremen is advantageous, especially when turning right to the east above the Brook?

In the past years, I had already made some surprising observations, which I have described in the previous chapters. But now, I was confronted in a completely new way with the question of the communication abilities of the cranes and their emotional world. These were questions that would not let me go.

Breeding Season: A Tragic Love Story

"Romeo and Juliet" were dear to my heart. Their joy when they came back to their territory will remain in my memory for a lifetime. I visited them several times every weekend, because I wanted to know if they would breed again in the "Alder Swamp" (a small, wet, wooded area with alder trees). (More than 20 years later, I was to live only about 300 meters as the crow flies from this little patch of wet forest.)

Two weeks after their arrival, I actually discovered the birds there in the small wet location (again, from a safe distance from the edge of the sanctuary, so as not to disturb them). They were building their nest.[1] Despite my good spotting scope, I almost missed it; who knows how many times I had missed them! The crane pair were moving at a slow-motion pace, so it was hard to make them out, despite the loosely standing trees. It was only because a heron flew over the alders and one of the mates gave a short warning call[2] that I had recognised the cranes—perhaps he had suspected a "hostile" crane but then realised his mistake, so "begun but not finished" did this territorial defence call sound. Two weeks after nest-building—I had not been able to observe any courtship or copulation—it became clear: "Romeo and Juliet" were breeding. They no longer appeared together, but each on their own; they no longer lingered in the meadow where they had usually been seen as a pair. I have often observed that pairs come to prefer a particular meadow before breeding, and then—during breeding, when they forage alone—often feed outside or inside of the brook in a completely different meadow or clearing, with each of the partners mostly choosing to feed in a different place. After the juveniles have

---

[1] One of the two cranes in the "Alder Swamp": https://www.bernhard-wessling.com/Erlenbruch
[2] cf. https://www.bernhard-wessling.com/warnruf; call record file: http://bit.ly/2MJZ4zi

fledged, they again seek out other areas—this then depends very much on the timing of the harvest, what the local farmers have grown in the fields and when hay was made. So, there are seasonally and individually influenced habits here, as well as changes in routines. One morning, I crept very carefully along a farm track near the small swamp and noticed: One of the cranes was sitting on the nest.

So, they had built their nest long before the brood, very secretly and carefully, and then, for 2 weeks, they gave the impression that they were up to nothing at all, that they had no nest or had it somewhere else. They had to feel that they were being watched. Of course, I myself came into question (although I had not entered the area), but above all were the farmers, hunters and foresters who drove and walked close to the nest in the vicinity and on a farm track, apparently without noticing the crane brood.

The cranes had to be wary, especially of foxes, martens and crows. In this respect, their caution was understandable. And at first, it seemed to bring the desired success. For 2 weeks, care of the brood proceeded normally. Disturbances by the local hunter (who, in my opinion, spent far too much time in his hunting district) or the farmer who looked after the hay meadow seemed to be stressful for the pair (at the approach of a human, a crane will get up from the nest and leave it in flight, but crouched, presumably in order not to reveal the location of the nest). In cool weather, there is a danger that, if a disturbance continues over time, the crane will not sit on the nest for too long and the eggs will freeze. There is also the risk that crows will take advantage of deserted crane nests. But this brood continued to be protected day after day.

One evening, a friend who was on crane guard called me: "I think they've aborted the brood." He described to me that "Romeo and Juliet" had been seen together out in the meadow for quite some time and no longer seemed to be tending the nest. I was very concerned and decided to put in two extra observation hours early the next morning before work.

By the time I arrived at the territory by foot, I had already heard strange calls from far away. Since I did not want to scare away the cranes that might still be sitting on the nest, it took me a long time to get to the place on the farm path from where I could see the nest with my telescope without disturbing the cranes.

The nest was empty. But close by, I could hear the unfamiliar calls of a crane that seemed to be flying, for the sounds were sometimes nearer, sometimes farther away. I felt at once that something was wrong. For a while, I couldn't see the calling crane, because it was flying at a low altitude above the nearby forest, so that the trees obscured it, and it was still early dawn.

Soon, however, the crane came closer, circled over the "Alder Swamp", over the adjacent meadows and called again and again—as I abruptly realized—in a manner I would describe as "plaintive" and searching. I knew immediately: His mate was dead. I had already read that when their mate perishes, by accident or otherwise, cranes will continue to try to find their dead companion for a long time, searching by flight and issuing mourning calls. What couldn't be adequately expressed in the books was that those calls would go right through your spine. I had no certainty at all, I saw nothing and so far had only spent a few minutes in the territory, and yet the mood of the flying crane was immediately transmitted to me. I mourned the lost crane. I "knew" she was dead. The friend had told me the night before that he had heard gunshots— not many, but scattered. Who had been shooting here in the spring? And why?

In the evening after work, I returned to the location. I continued to watch cautiously from a distance, still hoping that I was wrong and that the brood was still extant. It wasn't until 2 days later that I decided to go to the nest. No crane was rising. No calls could be heard, except for the mournful search calls. The nest was neat, no egg shells, no signs of fighting, but there were boot prints by the nest. Once again, humans had stolen crane eggs, the second time in 2 years. And I was convinced that the dead crane was the work of these poachers. In addition to two eggs, a stuffed crane, our "Juliet", was now to decorate a rustic hallway.

That very evening, and on many subsequent days and nights, I examined the surroundings of the "Alder Swamp". Perhaps the crane had fallen victim to a heart attack or a fox (both very unlikely) or an eagle owl (not quite so unlikely). If so, I'd have to find feathers or bones. Besides, crows would show me the way, for they would perch on the remains. I found nothing of the sort.

Each time, however, I encountered the surviving mate, "Romeo." He immediately proceeded to change his calls when he was on the ground and I was walking around his territory very slowly. He called to me. These were soft, rather desperate and accusatory sounding, "rolling" tones[3] that I tried to return. After a few days, I managed to get him to answer when I called him.

In the meantime, he had become completely lethargic. He hardly foraged any more, at least not when I watched him from a distance, and certainly not when I was nearby. He flew only his search rounds, which covered larger distances in the brook, always with the same calls. But within the territory, when I was near him and searching around, he let me approach him, up to 50 meters and even less (this is very astonishing, if you consider that the normal flight distance of cranes in Central Europe is about 300 meters). If I got any

---

[3] Example of "rolling" or "cooing" sounds that I recorded later: http://bit.ly/32QcYFG

closer, he would do some listless steps or wing beats and settle down only a dozen meters away. As soon as I sat down, he stopped, looked at me, and frequently called to me in short bursts. I called back. Exactly what I said then, I don't know (then, as now, I don't understand "Cranian" except for a few bits of "vocabulary"). It must have been all right though, at least, he didn't fly away or attack me. Despite an intensive search, I still found no feathers or remains of the dead crane. Finally, I gave up. The widower stayed in his territory until mid-June and then disappeared without a trace.

During the same period, I observed how the neighbouring pair, the "Sly Dogs", gradually took over his territory, proceeding very gradually and cautiously from the neighbouring "Green Brook" (I reported on this in chapter "Problem Solving, Ballet Courtship and the Fox Alarm: How Do Cranes Communicate with Each Other?"). "Romeo" had no more strength nor interest in defending his territory.

That's how "Romeo and Juliet" ended. What became of "Romeo", I do not know. I couldn't recognize him and don't know if he was the same crane that showed up in various territories the following year in February/March that we called the "Disturber". This crane seemed to fly into territories everywhere with the sole purpose to disturb. Maybe he was trying to lure a mate away and win her over? I don't remember when exactly I decided that I wanted to learn to recognize cranes individually—but the experiences with "Romeo and Juliet" were definitely an important impetus.

One March day, I observed a single crane flying north (there is a beautiful, large raised bog where cranes also breed). After that, we did not see the "Disturber" again that year. But in 1997 and 1998, we again had a strangely behaving loner who approached other pairs each time—perhaps the same bird?

One may criticize that I read too much emotion into these incidents. But it is the case that I myself feel something (that is: I care about what is happening) when I watch the cranes. But it is not only that: I mean to recognize that the cranes themselves have similar feelings. I know this may be considered "unscientific", because I have no proof. I would never allow myself to make unproven claims in my ancestral subject, chemistry, but here, I would like to express this opinion, even though it is itself based mainly on a feeling.

I have come to doubt more and more that it is right to deny feelings to cranes and other animals. I believe that this does not do justice to the complexity of these creatures. I am convinced that many animal species—like the cranes "Romeo and Juliet"—have a spectrum of feelings such as (life) joy, amazement, boredom and sadness. It would not surprise me if these feelings communicate themselves internally to these animals differently than they do so to us (I assume that each of us experiences feelings differently, and people from other cultures experience their feelings differently than we Europeans do). We will deal with this question in more detail in chapter "What Can We Learn About Intelligence, Migratory Behavior, the Formation of Culture, Tool Use, and Self-Awareness in Cranes?" and in the appendix, and we will see that my observations from that time are quite consistent with modern scientific findings about the role of emotions in humans and animals, including in light of evolution.

Ruffs, White-Tailed Eagles and Other Brook Visitors: What Crane Guards Can Experience

My experiences with the crane observations in the Brook have shaped my image of cranes. Of course, most days during a crane guard week were rather "normal" and, from an outsider's point of view, would not have been particularly spectacular, or, indeed, may have seemed rather boring. However, I always felt that the "normal" observations were interesting enough in themselves.

Normally, my son and I mostly rode our bicycles separately in different rhythms along the paths and to places from which cranes could be observed. In addition, we looked at the locations where visitors often deviated from the path. What we saw was usually "quite normal": cranes foraging or grooming their feathers, a green woodpecker, a fox, walkers on the paths (and not in the bush). When we met again after a round trip, we reported to each other about what we had seen.

But shortly after Easter 1997, in the same week when it was Bengt's and my turn as crane guards, more happened in one day than one sometimes experiences in a whole year.

When I got up at quarter past four, Bengt was already on his feet. The morning was clear, it was pitch dark, and only the moon helped us to find our way. The comet Hale-Bopp was relatively visible at that time, considering our limited viewing conditions in Northern Germany. None of this could compare with my observations a few days earlier in Iceland, on a starry, ice-cold winter night in an extremely lonely, even much darker area: I had first gone there on business the previous February, because Iceland was ideal for testing my new corrosion protection technology under real conditions. After that, I spent a few days in a wooden cabin in northern Iceland. I often took a few days off near Akureyri in the north of Iceland after the technical meetings in

B. Wessling, *The Call of the Cranes*, https://doi.org/10.1007/978-3-030-98283-6_6

Reykjavik and Akureyri, because I was able to sort out my thoughts in the solitude of long walks and write patents and publications undisturbed in the hut in the early darkness. On February 25, there was a very clear night there at Eyjafördur, and I noted breathlessly in my diary afterwards:[1]

> Cold, -10°. Hale-Bopp, the comet ca in the N, maybe 20° high, long tail (1 thumb wide!) already visible with the naked eye, but even more with binoculars! I stay outside, marvel at Hale-Bopp, aurora + stars. + Andromeda. Auroras: wide bands in arc from W → E, 35° to 90° high, partly stable for a long time, more like backlit cloud edges, but then fraying, into swirls, filaments, merging into everything! Crazy!

In the Brook, I now again had some time to observe the comet for a prolonged period, although it was much less spectacular.

I had taken my place in the darkness at the edge of the "Alder Swamp". From there, I had, on the one hand, a good view into the area, while on the other hand, I was far enough away not to disturb the cranes during the time when I would need to retreat at first light.

It was cold, five degrees below zero. I heard the snipe "grumble"; there were two of them making their territorial flights. Every few minutes, a woodcock flew by, which I could also only hear but not see. On this morning, there were two birds apparently courting. They were louder than usual.

Gradually, dusk began to fall. The next birds to make themselves known were the raven crows. A song thrush began the morning concert of the songbirds. Then, a loud, clear and long unison call of cranes from the "Alder Swamp" area. Silence. A crane pair from the "Green Brook" answered. Oh, I thought, that's new, since when is there a pair here again? I took it upon myself to keep an eye out for another crane pair trying to settle in this obscure territory. If this was the case, it was probably one of the "youth groups" that had disbanded a few days ago.

Another unison call from the breeding area, a deep but small waterhole, surely a former peat pond, which, at that time, was already surrounded by bushes and partly silted up. Now, I tried to keep an eye on the nest area, because I wanted finally to know how the cranes would manage to hide their

---

[1] I kept a diary during one of the most exhausting, critical and stressful professional/business periods of my life, the decade from 1996 to 2006, though not daily, but rather every few weeks when I "felt" a phase had been completed, a step forward had been taken or a massive setback had to be dealt with; I usually wrote when I had some peace and quiet—on the plane, in a hotel in the USA, Japan, or Korea, or when I spent a few days alone out in nature or went diving, usually after a period of intense business activity somewhere in the world. This diary also contains many notes about my crane research.

nest this time. So far, I only knew where they were breeding, but not how they flew out of the nesting area.

A quarter of an hour later, I observed a mate moving to the edge of the forest. The nesting area was very damp and only loosely covered with trees and bushes. I could see with my binoculars—from a distance of about 500 m— certainly 10–15 meters into it. Nevertheless, I could not see the nest itself. The crane moved closer and closer to the edge of the forest as if in slow motion. But then, he moved to the left, a few meters away from the open edge. Now, with a slowness that was almost torturous for me as an impatient observer, he moved farther and farther to the left, along the inside of the forest edge.

About 150 m from the nest, he stepped out of the woods, suddenly moving quickly, and immediately flew off, flying at a height of only 2–3 meters to a nearby meadow, where he began to take his breakfast. I remained sitting. The redwing thrushes and a few starling groups appeared and made a merry racket. They sought their breakfast in front of me in the meadow.

After 45 min, the crane suddenly swung up, flew south and landed, no longer observable by me, behind the nesting area. Ten minutes later, I heard the unison call—ah, an exchange about breeding responsibility. Another 10 min later, the other partner flew off to the south from the back of the wet woods. The crane pair hid their nest quite cleverly with such strategies.

I left the area and made my long round tour. Bengt and I radioed news to each other now and then. He told me about an encounter with a wild boar that had snorted past him that morning at a distance of only 10 meters. He said that it had given off a very intense smell, which was his nice way of saying that it stunk.

I passed by a meadow in front of a moor birch forest patch, where a crane was having a hard time with a lapwing that was attacking him. What was the reason for the lapwing to attack the crane who was innocently searching for food while its partner was on the nest? Perhaps, the lapwing wanted to establish a brood on this meadow, or its partner was already breeding? I did not find out, but I was happy to see a potential new lapwing territory (Fig. 1).

After our second breakfast, Bengt drove to the previous "guard spot" (as we called a particular place in the Brook where we crane guards were, for many years, able to spend the night in a caravan that was temporarily parked there during the breeding season, until the nature conservation authority forbade this, after which we would spend the night in the forestry depot). I drove along the "Old Stake Trail", from which one can look from the south into the "New Moor". I hoped to be able to stay there for a while, so I unfolded my fishing chair and made myself comfortable. Bengt and I wanted to see if we

Fig. 1 Common Crane tries hard to keep the lapwing at a distance. (© Bernhard Wessling)

could coordinate our observations from the north and south and learn more about the new crane territories.

It was nine o'clock by then, and the Brook was quiet. The lapwings in front of me were mewing. Suddenly, within seconds, all hell broke loose: Ducks, geese and herons were in the air, screaming excitedly. Bengt radioed: "There's a white-tailed eagle here! It's hunting, it may even have caught something, it's hard to tell." Ah, hence the excitement. Many herons were standing on the very tops of the trees; it looked funny, because a strong wind was blowing and the herons were swaying—but they probably preferred that to being at the mercy of the eagle.

Over the radio, I heard, "I think he's got a water rail, but I can't see him, he's flown off to the back, can you see him?" Unfortunately, I couldn't, so the white-tailed eagle had probably disappeared between the reeds and the group of birch trees. Slowly, silence returned. But we were thrilled, because, for the first time in a long time, a white-tailed eagle was our guest again.

When we passed the information centre, a visitor asked us if it was possible that it was a white-tailed eagle that was perched nearby. Yes, of course, so he had flown here! And there he was, sitting on a dead birch tree behind the information centre. We showed the eagle to the woman's children through our spotting scope. They were impressed: He was so close, filling the spotting scope's entire field of view. He sat there very still, hardly moving, just turning his head sometimes to the right, sometimes to the left. During the course of the day, we took several quick looks—he continued to just sit there. Apparently, he had caught and eaten well and was digesting in peace.

In between, I went back to the previous guard spot. Due to our attempts to find the eagle again, the cranes had, for a short time, no longer been the focus of our interest. That changed quickly. I saw a crane up ahead in front of the reed belt. A single crane? That was unusual, but I remembered the previous evening: There had been another pair that had invaded the territory of the resident pair, the "Foxes," while they were still out for dinner. The young pair had been separated during the late twilight territorial defence hunt, one of the partners remaining at the front of the reed belt. Had he stayed there overnight and not stirred until now, in the middle of the morning?

Now, he came to life, as a short, monotonous call sounded from beyond the reed-belt, a call that I recognized as a search and contact call. He raised his head and listened. Another contact call. And another. He took a few steps, and then called in turn. A single search call, along the lines of "I'm here, is that you?" From behind, the reply, "Yes, it's me, is it you there?" Another search call there, one back, one there, one back. Then, the single crane soared, flew over the reeds and landed. A joyous duet call rang out, "There you are again!" For the first time, I witnessed the reunion and recognition of a crane pair that had been separated (if only for one night).

In front of us, a Ruff landed. One of the employees from the information centre had already seen it in the morning and had told us. It was unusual that a single Ruff, for half a day, maybe even since the day before, would be resting here on its migration. Now, something strange began to happen. At first, it seemed as if one of the Lapwings (one of four that had their territories quite close to our observation site) wanted to chase the Ruff. Strange, because it had been allowed to sit there and look for food for some time—what had suddenly changed? The Lapwing flew in on an "attack". The Ruff flew up, but instead of fleeing, it made turns, loops, and roller coasters, the Lapwing always beside, in front of, or behind it. There was a merry game going on here, involving two birds of different species—it looked like pure joie de vivre.

The Lapwing encouraged the Ruff to perform aerobatics several times during the afternoon, and the latter joined in again and again, but with less and less stamina over time. While the initial flights lasted for a whole minute or even longer, the Ruff later began to set down in the meadow after only 10 seconds. The Lapwing did not like that.

He now attempted to enlist one of the many snipe that had been squatting there in the grass for a few days in his game. He shouldn't have done that, because he couldn't keep up with the snipe! It flew up, zigzagged flat to the right, left, right, straight ahead—and was 30, 40 meters away in seconds, while the Lapwing only wanted to do aerobatics! Disappointed, he set down in the grass and let the other birds hop and fly their way.

Hadn't I observed something very unusual here? Birds of different species playing with each other? It was clearly not a fight about territory or food; rather, these flights seemed to express the pure joy of life. "Interspecies play"— I didn't know that such a thing could exist, except between humans and dogs or dogs and cats that are friends with each other—but between wild birds? Very amazing, at least to me!

Bengt and I had planned to spot the first osprey that day. Ospreys had not bred in Schleswig-Holstein or Hamburg for decades. Therefore, we only saw them when they passed through our area and sometimes stayed for a few hours or days. After we had already seen the white-tailed eagle, our desire to see an osprey now seemed almost a bit impertinent. Nevertheless, we pointed our binoculars and telescopes at any bird of prey that became visible in the distance, because it might be an osprey. But they always turned out to be just brightly coloured buzzards.

When Bengt came back from his round, I went off, as we were taking turns. Fortunately, he had not met any "area violators", i.e., people who wander off the trail and into the closed areas; maybe I would be spared them that day as well.

I passed a spot from which, with a little skill, a crane could be observed on its nest. The pair in question was always in this location, but we found that they were breeding a bit boastfully. We took extra care to make sure that no photographers there felt motivated to get even closer to the nest. After all, you could already take nice photos of the cranes from the path with a medium sized telephoto lens.

On this day, a special spectacle presented itself. The white-tailed eagle, which I had thought would still be on the dead birch near the information centre, was now perched on a tree stump not 20 meters away from the crane's nest. One crane was sitting on the nest, the other was standing vigilantly 5 meters away. I waited. After some time, the eagle began to move; it soared and made a few circles over the crane's nest. Immediately, the two cranes called out to each other; the one perched on the nest stretched its beak aloft, ready to defend itself, but didn't budge. Her partner moved closer to the nest and also stretched his beak into the air with his neck slightly back, ready to hack at the eagle's head or belly from below if it dared to come closer.

The eagle was certainly well-advised not to attack the cranes. He seemed to think so too, and subsequently moved off into the distance, where I could no longer see him. I drove on.

When I arrived at the "Little Moor", which lies at the edge of the Brook, I saw a car that had apparently just stopped there. A young couple, "smartly" equipped with hiking boots, got out of what we now call a "SUV". I frowned and quickly cycled there, suspecting they were up to something illegal. And sure enough, they bypassed the trail block, intending to go through the hedge line into the closed portion of the Brook. I approached the two. "Just a moment, excuse me, where is it that you are planning on going?" They wanted to hike there, as they thought was obvious—why did I ask? And who was I, anyway?

I am always, at first, very friendly, but at the same time determined, and at this point, I wanted to let the two understand that under no circumstances were they to enter that part of the Brook, that the path was closed for a good reason: For a few days previous, the "Little Moor" had been visited by cranes that perhaps wanted to occupy a territory there again.

We debated back and forth—a little frostier at first, then a little more heated. I could not be persuaded. Finally, the man burst his collar and indignantly stated: If they were not allowed to enter the brook there, then they did not want to hike in this area at all! The two then turned on their heels, got into the car and drove away. What a loss for the Brook!

From the "Little Moor", I heard crane calls. I drove on. At a trail junction, I saw two bicycles standing there; that alone was suspicious: Where were the cyclists who belonged to them? I looked for them and found a father with his young son, who was shouldering a wooden rifle. They were just heading into the woods. "Wait, excuse me, where are you going?" That was a stupid question, the father said, they were here to play a little Wild West, why was I bothering them? "Because this is a nature reserve where it is forbidden to leave the paths." Again, I ended up having to argue for a long time. The father was, as he described himself to me, a doctor of biology and claimed that he knew exactly how important nature conservation was, but also that one didn't have to adhere so exactly to such rules. Besides, he said, this was the only area far and wide where dogs were forbidden and where one could play without constantly having to step in dog poop. He was getting increasingly angry: What kind of state was this anyway, where you either had to step in dog poop while playing Cowboys and Indians or were expelled from the forest by overzealous conservationists?! Unfortunately, I couldn't help him, neither could I tell him where, outside of the nature reserve, there were places with no dog poop, nor could I give him special permission to go off the beaten track and play Wild West in the nature reserve.

After this unpleasant discussion, I drove on, to the south of the area. In that place lies a "recreational forest", which is under some nature protection, but is more of an urban forest than a nature reserve. You are allowed to walk dogs there if they are on a leash. Today, however, I was pursued by bad luck, because I saw a couple with a big dog, which, although leashed, was pulling them towards the entrance to the Duvenstedt Brook—that was, unfortunately, the wrong direction. I overtook them and waited on the bridge over the small Ammersbek river, right by the sign that said, "Dogs forbidden, even leashed".

Things progressed just as I had assumed they would: The couple, who, dressed in noble threads, made quite a civilized impression, wandered unconcernedly past the sign into the nature reserve. I called out to them, asking them if they might have missed the sign. "No, what sign?" They then admitted that they knew of the sign, but claimed that there was no mention of dogs on it.

We went back together and read it together. Now, the third debate of this day started. They didn't want to go all the way through the area, but only along such and such a path, their dog listens very well to them, is leashed, doesn't cause any disturbance at all, and all of this was just harassment. Between all of these encounters, I was beginning to feel like a traveling preacher, but that couldn't be helped. The simple fact was that the dog had to stay outside.

Somewhat exhausted, I made my way back to the guard spot, from which my son was already radioing, asking me where I was. We wanted to eat something and then enjoy the evening, hoping for the flocks of starlings, which conducted their aerobatics over the Brook. A man and a woman approached me. I had already met them at the observation site and talked to them briefly. They wanted to know more about crane protection. They could well understand that there was a lot of trouble with people who had no comprehension about the needs of the area and just did whatever they wanted to do, but could one still have fun there? "Yes," I replied, and told them about the white-tailed eagle. They had seen it, but had not recognized it. I told them about the crane breeding successes, about the problems we had with stolen eggs, for example, and I mentioned the starlings and their roosting flights. They were really interested, and wanted to come more often in the future and ask more questions, because the more you knew, the even more interesting and enjoyable the visit would be. They had already questioned Bengt: "What, that's your son standing up there right now? He knows a lot about birds. It's interesting that you, as a professional in a completely different occupation, invest so much time with your son and for nature protection." I was pleased, because this meant that I had had at least one positive conversation with the Brook's visitors that day, too. Fortunately, such encounters happened all the time.

After a quick meal, my son and I stood at the guard spot, from which one can get a very good view of some meadows and the inner moor. A large troop of snipe flew in. The Ruff was teased by the Lapwing, but wouldn't fly with it again.

Slowly, it became evening. The first starlings came. They sat down on one of the large chestnuts that stood by the forestry operations center. More and more arrived during the next half hour. They "chatted" in confusion, then were suddenly silent and flew off almost all at once. First, they formed a thick cloud, then a tube, then a "pear", which seemed to be magnetically attracted to the chestnut—and then they all landed again.

This preliminary banter was repeated in variants a few times. Even during the flight, they chirped, but with different sounds than after landing on a row of trees, after which they were suddenly silent again. This went on before the real spectacle began. As if they had waited until "all of them" were there, the starlings flew in immense clouds, tubes, rings, forming carpets and whirlpools of bird bodies, swirling about and rearranging themselves into ever new, mobile and infinitely changeable formations. In those days, we saw by far the largest gatherings of starlings in recent years that any of us could remember. Bengt and I tried to estimate how many there might be. We tried to do it with the geometry of a tube that formed: "This is now a complete ring, diameter about 400 m, that's a circumference of $2\pi r$ ("two pi er"), so about 400 times pi, which makes more than 1200 m."—"And the cross-section, if it were square, estimated to be at least 20 by 20 meters, equal to 400 m², times 1200 m in length, which makes 480,000 m³." We were astonished, for if there were now only one starling flying per cubic metre, it would have been nearly 500,000—ten times more than had been estimated there a few days earlier.

At times, this flock would stop near the row of oaks to our left, land there to chat briefly, and then fly up again, only to land immediately afterward in the meadow in front of us. There they sat for only a few seconds, chatted, went silent, rose a few feet, and then landed in a compact group, having advanced ten or 20 meters closer to us. This happened a couple of times, until the starlings landed together only 20 or 30 meters in front of us and rose again, nearly sweeping our caps off our heads with their combined gust of wind.

Now, we could better estimate the number of birds; because their formation marked a roughly square area close to us, we could estimate its size in square meters and count the starling density per square meter. We came up with about 250,000 animals—an impressive amount. I was fascinated by this phenomenon: How did the starlings, who came from all directions, know that they would meet so many of their kindred in the Brook? How did they communicate the meeting place? Why did they meet in the evening, and what did these fantastic roosting flights mean?

At the end of the natural spectacle, they flew over the moor again, drew some figures in the sky for a long time, but then suddenly vanished as if sucked within seconds by a vacuum pump into the moor reed, where they spent the night.

But the day was still not over, and it held another spectacle for us: We were able to watch for a few more minutes as the white-tailed eagle circled right in front of us, 200 meters away, hitting the water again and again, and finally, obviously, snagging his prey. And instead of darting away as he usually did, he perched on the stump of a fallen tree and began tearing into the victim. We looked through our spotting scope and thought we noted that he had captured a water rail. His supper kept us occupied until well into the twilight. When do you get to see something like that?

While the white-tailed eagle was eating its prey, two cranes flew in from the east and landed nearby. Two? I wondered. The resident territorial pair, the "Foxes," were already breeding there, meaning two additional cranes would not be allowed to fly in. And sure enough, 10 minutes later, another crane flew out of the area where we suspected the nest to be and came down firmly next to its two conspecifics, not far from the placidly dining white-tailed eagle. When the intruders didn't seem to understand his message, he didn't hesitate long, lunging towards one of the mates and scaring it away.

It became a turbulent chase. The two squabblers flew for a long time along all possible curves, after which the second intruder also took to the air. It probably didn't want to lose track of its partner in the twilight. Eventually, the two intruders flew away. The territory owner returned, stood briefly next to the eagle, which continued to calmly dismember its prey, and then slowly walked back in the direction of its hidden nest, where its partner had patiently continued to care for the brood.

What a spectacular ending to a momentous day! In the meantime, it had become completely dark and we could no longer see the eagle, even with our good binoculars. There had also been nothing to see or hear from the starlings for a long time. Bengt and I decided to call it a day as well and go to sleep. With all of my many crane watches, there was probably no other day as eventful as this one.

In the School of Life

Some behavioural biologists or (animal) psychologists might regard my observations as unscientific and as a worthless collection of "anecdotes". I am well aware that I have not performed any verifiable and repeatable experiments and that I had no experimental designs. Therefore, my observations cannot meet the criterion of "repeatability". My observations were not made under controlled conditions, they are not accurately recorded (for example, with tally sheets or video recordings), except for short notes and early drafts of chapters for this book, and only some of the incidents have additional witnesses.

In my opinion, the observations nevertheless have enormous value, including in scientific terms, because, like most events in our lives, they are unique, but at the same time "true", even though they defy the logic of a repeatable experiment. They provide a background against which new hypotheses are developed, which can then be tested with the help of controlled scientific experiments.

If one examines the cognitive performance of animals, one is always faced with the problem that, although one can proceed "scientifically precisely" in the laboratory, this procedure is certainly not in line with the normal life of the animal in question. Why should a wild pigeon distinguish triangles and squares and count the number of points? On the other hand, we can draw tentative conclusions as to what mental capacities the pigeon has. However, the cognitive performances that animals perform in the wild cannot be determined in this way.

If, instead of laboratory experiments, we make field observations without disturbing the animals, we can, with some luck and skill, observe how they behave in "real life", what real problems they cope with and how they do so.

© The Author(s), under exclusive license to Springer Nature Switzerland AG 2022
B. Wessling, *The Call of the Cranes*, https://doi.org/10.1007/978-3-030-98283-6_7

In such observations, we should not be experimenters, interfering in any way with the animals' lives, for example, by putting out food or changing the environment. This means, of course, that we have to be much more patient, that we cannot set up targeted experimental designs, but have to wait and see what questions nature poses to the animals and how they respond. And we have the disadvantage of not being able to do control experiments and only very rarely being able to achieve the scientifically required repeatability. The advantage, however, is that we observe life as it actually happens.

Perhaps, in the eyes of critical readers, I have sometimes been a bit too brash in my interpretation of the cranes' behavior. For all those for whom my previous conclusions may contain too much interpretation, I would like to report below on a few observations that do not seem so exotic and yet clearly describe the cranes' ability to learn.

Anyone who has ever observed cranes, for example, during the breeding season in sparsely populated areas of Sweden or Brandenburg or while wintering in Spain, will have noticed: They react attentively and uneasily, even from quite a distance, when people approach, and they warp or fly up as soon as people get closer to them than about 700–1000 meters.

In Mecklenburg-Vorpommern, but especially here in Duvenstedt Brook, the cranes behave differently. They are, to some extent, accustomed to humans, have established nests near settlements, but still behave very shyly.

This can be easily observed in the Brook. In some places, the paths run close to feeding places that the cranes like to visit. Here, it can happen that cyclists and sometimes hikers are tolerated by cranes, even if they pass by at a distance of only 100–150 meters. The cranes often become restless when a cyclist or a large group of hikers stops to watch them. They usually fly off.

Apparently, at least some cranes have become accustomed to humans as long as they stay on known trails. Similar to red deer and fallow deer, cranes know exactly where the trails are and when they can assume that humans will leave them alone. For example, where we had closed a trail, the cranes quickly learned that their territory had increased as a result and they conquered the zones to the right and left of the trail in question, as well as the trail itself. In general, "our" cranes have a flight distance of 250–300 m; if they are guiding chicks, about 500 meters.

However, we can observe very different behaviors. The "Foxes", for example, were not at all shy on their preferred food meadow far away from their territory. I almost don't dare to report that they let people get as close as 25 m early in the morning. I have experienced this myself, as have other crane guards. We accomplished this by staying on a main path, and were even allowed to stop.

However, while cranes sometimes tolerate people in close proximity on the feeding meadow, they keep their nesting area strictly hidden from human eyes. Only the "Pioneers", in 1998, chose a nesting site that was visible 200 meters from a path, and that was very well shielded by water all around. Unlike other pairs that make detours when approaching the nesting area to keep it secret, they almost provocatively flew directly into the nesting area and offered unison calls when exchanging the brooding responsibility, which made us crane guards almost uncomfortable. Here, visitors to the Brook had a crane's nest right under their noses, if they had the patience to look for an albeit narrow window of view of the nest between the densely standing downy birch trees.

The pair that bred in the "Little Moor" for many years, the "Townies", led their juveniles to within 100 m of a passing road—but never on weekends after nine or ten in the morning. Almost all cranes can be observed getting closer to paths early in the morning than later in the day; on weekends, they retreat much further, no doubt because of the many visitors. Just recently, when we went on another walk into the Brook directly from home, we were able to observe a family of cranes with a juvenile just 30 meters from a main path. We had encountered no other visitors to the Brook that morning. A few days later—there were again no other visitors in the Brook due to bad weather—we encountered another crane family with a chicken-sized chick just 5 meters off the path. So, I guess it's safe to assume that the cranes are very observant of their surroundings: When the paths are at least occasionally used by walkers, they stay 200–300 m away from them; in bad weather and very early in the morning, they go right up to the edges of the paths, sometimes even crossing them because there are food sources to the right and left of them that they could not otherwise use. In regard to us, the chick carelessly playing so close to the edge of the path allowed us to walk on very slowly, breathless with wonder, stopping again and again. We had never before been able to observe a crane family from so close and for so long; as I had not brought along the big camera with the telephoto lens because of the uncertain weather, I photographed cranes for the first time with my mobile phone.[1]

The "Sly Dogs" had, as already described, very efficiently looked for new areas—in addition to the territory of "Romeo and Juliet", in which no cranes had been previously seen. They lost their former shyness. Now, they suddenly began standing in meadows close to paths, one day bringing a cheerful (or combative?) duet to the ears of one of our crane guards from a distance of 10 meters and once again moving their breeding place. With them, I experienced

---

[1] cf. https://photos.app.goo.gl/kWbiium6ii6ENxgp8

a pattern of behavior that I had not seen before: After a while, they foraged regularly—every day, really—in the mornings from about seven o'clock on, in a very specific meadow on the edge of the nature reserve. In the afternoons, they flew farther out and could be seen in another meadow close to the village, in the neighbourhood of a horse paddock. Why this temporal pattern? Why did the "Sly Dogs" need multiple foraging areas at all, when other pairs got by with just one? I think that the birds were looking for variety, that they felt a certain curiosity. Maybe they liked to eat something different in the morning than in the afternoon.

I also had such thoughts when I saw that one of the "Pioneers", during breeding, repeatedly left the territory in the evening at advanced dusk and flew to a food meadow ... but did not look for food there at all, choosing simply to "stand around". One of the "Foxes" behaved similarly. With other pairs, I was not able to observe a comparable behaviour.

New pairs are insecure at first. When one of the partners is breeding, the other one does not fly far away; it probably wants to stay within calling distance, presumably to be able to fly back quickly "in case something is wrong". The "Seekers", for example, behaved particularly cautiously in their first breeding season in 1998: From the beginning of the brood, they hardly called at all and were not to be seen in the pastures that they had visited before laying their eggs. As the breeding season progressed, however, the cranes would occasionally visit other areas they knew from before they laid their eggs. With growing experience—I noticed this especially with the "Foxes"—the partner not sitting on the nest would fly directly to the best meadow on the first breeding day, farther away, presumably without worrying about anything.

In 1998, we observed a crane pair foraging before nine in the morning up to within 100 meters of a path. Then came the first of May. This date put all members of our group, who assist the crane guards on weekends and holidays, on heightened alert. Because all too many people from Hamburg, as well as the neighbouring district of Stormarn, were running around in the meadows, woods and clearings, picnicking, and playing football, Frisbee or basketball. They used the nature reserve as a recreational area and disturbed the breeding birds (not only the cranes). Apparently, the cranes found this day to be as stressful as we, their guards, did, because, on days like this, they flew around but were not found in the usual foraging meadows, choosing to remain deep in their territories instead. The pair in question, however, was to be found, on that first May evening after half-past seven, in the usual vicinity of the path. The May Day visitors were barbecuing in their gardens at home at this time.

So, it is clear: "Our" cranes have learned something that other cranes (for example, in Sweden) do not have to learn. They know that humans come into their environment (mostly) on paths that delineate certain areas in which they themselves can then move safely (and people have also learned through our conservation work: the Brook is mistaken for a city park meadow far less often than it used to be). Some cranes may deduce from this that humans are generally harmless. However, the crane pair whose eggs were repeatedly stolen by humans in 1998 again have hardly been seen since. It has retreated very far, perhaps because it has learned something else?

Crane observers and researchers are increasingly of the opinion that cranes in general have learned, in recent decades, to take poorer breeding sites and to move closer to settlements. This fits with the behaviour of the cranes that we observe here in the Brook.

The ability to learn varies from individual to individual, depending on experience (and the willingness to learn). This can be seen from a simple observation: Cranes seek out specific areas in which to feed. When they fly there from their territory or back again, they always take the same route. They fly in a direct line to their destination, usually at a relatively low altitude. Perhaps this seems normal to you, but only if you consider how you move around in your own "territory", i.e., your home town. There, you take the direct route to the weekly market (only, because you can't fly, you have to follow the course of the roads with bends and junctions—you can't take the beeline).

But how do you get around in a city you don't know? Let's say you're in Tokyo and you don't have a map that you understand because you can't read the characters. Let's also say that you somehow found your way from your hotel to an interesting museum, but now you don't know how to get back. All you know is that it wasn't very far, maybe 15 minutes, only three or four blocks. You'll turn right sometimes, left sometimes, maybe make bigger and bigger circles. After a while, you'll find your hotel again.

The next day, you want to go to a restaurant that is right next to the museum. Since you have memorized a few prominent points, you can now already proceed on a much more direct route. After a few days, you know the district even better: You have created a mental map.

The cranes do nothing else. They no longer have to "circle" in areas they know, they fly directly. And that is anything but a matter of course. When approaching a territory, I can already recognize the territory owners by the fact that they fly in a straight line to their target area; newcomers to the area fly in curves, aimlessly and indecisively, often changing flight direction and altitude, similar to how you would move on your first day in Tokyo. And similar to how you would imagine your way to the restaurant on your second day, the crane will imagine where it wants to fly before taking off. It doesn't have to think in words when doing this—you probably don't either. You imagine the way figuratively, just as the crane does.

On an observation day before Easter, I noticed this once again: One of the "Foxes" came directly from his feeding area, the rush meadow, back to the territory for the breeding change, after which his partner flew without hesitation to the second pasture preferred by the two, the "hill view". Three hours later, there was another one on the rush meadow. This is typical behavior in many pairs of cranes: They fly directly and without detours, make different decisions, and thus choose—according to which criteria, we can only guess.

Once, when the "Pioneers" underwent a full moult after the abortion of a brood, they were unable to fly, and often were simply loitering around the edge of the reeds near their territory. There is a meadow with an old fence on one side. One day, they were walking across the open meadow towards the fence. I hadn't looked closely, but, in any case, the first one was suddenly inside the fence while the other one was still outside. Maybe the first one had jumped or could fly a bit, whereas the second one could not or would not jump over the fence. But the first went on and on, without stopping, without heeding the low sounds of his partner. The latter now walked along the fence like a caged lion, to the right, to the left, not knowing what to do. It took her a few minutes to come up with the solution: She walked far to the left—the fence makes a corner there—then walked along the side of the fence to the end where she again met her partner, who was able to pass under the fence there.

This solution was possible only on the basis of local knowledge. The problem had arisen only because of the moult, and finding the solution took quite a long time—but the crane found it! I see no other possibility: It must have mentally imagined the way around the fence before it actually walked it.

Another skill that seems so simple to us humans also probably needs to be learned and practiced just as much, namely, developing a sense of time. When cranes breed, they take turns. One sits on the eggs, the other searches outside for food. In the morning, it is often the case that brood changes occur at shorter intervals, sometimes as quickly as after half an hour. An average "shift" on the nest lasts about an hour or two. But at midday and in the afternoon, the foraging crane often allows itself longer absences. Three hours is not unusual then; I have known a crane to be absent for 6 h. How does a crane know how much time has passed? How do the animals divide up their day; how do they learn not to let their mate starve?

The observation of the young biologist from Osnabrück, which I mentioned in chapter "Problem Solving, Ballet Courtship and the Fox Alarm:

How Do Cranes Communicate with Each Other?", shows "active teaching": The crane sitting on the nest could not summon his partner, who had been absent for too long, so he admonished her: "Don't make me sit here so long again!"

I am sure that cranes also have to learn what food suits them, what is particularly nutritious, what tastes good and not so good. After all, almost all cranes grow up in different environments, and therefore have different food sources. Learning comes naturally here: In the first weeks, the crane parents feed the young. In doing so, they show them where to find something good and what that good thing is. This can be observed by anyone who watches cranes in the first weeks of their lives as they search for food together with their parents. After some time, the juvenile knows what is wholesome and what is less tasty. In the winter quarters, or even as early as the autumn rest, there are again new food sources and again something to learn (Figs. 1 and 2).

Cranes also have to learn other things that many people think are innate to them. Oddly enough, this includes flying, of all things.

Of course, the ability to fly is innate in cranes; after all, they came into the world with feathers, wings, and the muscles to go with them. This is comparable to the ability of human children to walk, but that didn't work out on the first day of our lives either; we had to learn it. Anyone who has watched a toddler take its first steps knows how difficult that is. Cranes don't just fly off

Fig. 1 A crane youngster, about 8 weeks old, gets some food from the parent bird, and so comes to learn what is edible. (© Bernhard Wessling)

Fig. 2 A crane chick guided by its parents on an opening in the moor birch forest patch learning where to find food. (© Bernhard Wessling)

either. So, in late July or August, there is a lot to see. This is when the young cranes learn to fly.

By this point, the juveniles have flapped their wings many times before without much happening. But one day (of course, we don't know if it's the first time when we're watching it), we see the adults walk in front of their offspring, flap their wings vigorously, and fly a few meters. The young ones try to imitate it, but it doesn't work. They jump, but they don't fly.

Now, one should not imagine that these first attempts look as exaggeratedly funny as one of the take-off attempts of the albatross with which Bernhard and Bianca fly in the Walt Disney film *The Rescuers*. I've never observed belly landings or anything like that. Watching young cranes make their first attempts at flight is nevertheless always touchingly funny and exciting.

"Look how I'm doing it," the father seems to call, flapping his wings in time with his steps, sometimes flying a few feet, then running again with powerful wing beats. The young ones again try to imitate their parents. And after some time, the first young bird finally takes off, but only for a few pitifully short metres, and no sooner has the take-off been accomplished than the next problem arises: How is it to land?

The whole body of the youngster expresses this uncertainty. While the adults stretch their necks resolutely forward at take-off, beak and neck forming a line, considerably elongating themselves and leaving the ground after one or two wing beats, the little ones have their eyes fixed rigidly on the ground, body and neck forming an angle, wings not stretched out in their full span—and yet, a successful sail jump still generally occurs. The young one then immediately lands again, for safety's sake, and runs out the remaining momentum.

Now, it flaps its wings while standing, strong and determined, but again, nothing happens; the crane's body remains on the ground, to the owner's disappointment. For the time being—and it will remain so for a while—taking off is only possible by jumping. So, more and more jumps have to be practiced, and at some point, it succeeds in flapping its wings on take-off and leaving them spread and bent forward for braking on landing. The way the parents are flying looks so easy, but the efforts of the young ones make it clear how difficult and exhausting learning to fly is after all.

According to the experience of the ICF, the legs are the most sensitive parts of the crane's body. A ligament can all too easily become stretched or even tear, or a bone may snap. The young cranes seem to be aware of this danger. Carefully and persistently, though not yet very elegantly, they jump off again and again and land by running out their momentum.

And suddenly, it becomes their very first glider flight! As if the youngster was frightened by its own performance, it does not fold its legs backwards in a relaxed manner, but keeps them stretched out in front, ready for landing at any time.

So, the flight school goes on, step by step, reminding me of the time when my children, and then my grandchildren, learned to walk. If we think that crane parents, unlike us human parents, can't explain anything to their children, we should keep in mind that human children can not speak at the time that they are learning to walk. What they understand from statements like, "Oh, how great, do it again like that, that was already really good; don't worry, you won't hurt yourself if you fall over!", we don't know. We communicate with our children not only through verbal language, but also through body language and eye contact. Human children learn a lot through imitation, and it is by no means self-evident to cognitive scientists that birds like cranes also learn through imitation. But they obviously do.

I've always wanted to see a crane chick take its first real flight. Unfortunately, cranes usually fledge at the exact same time that we are on holiday or on weekdays during working hours. In 1997, I finally got my wish, because one of the pairs had had to abort their first brood and had raised a very late subsequent brood. Their offspring was still not fledged in early September. High time to learn to fly!

I really wanted to witness the first real flight rounds this time, and therefore went to the Brook every day in the evening after work. The young bird was still not fledged. One Saturday, when I looked in on the crane family again, there were some flight exercises going on, but the youngster didn't seem to have the courage to take off yet. Again, the parents demonstrated how it was done, and the juvenile tried to imitate. To no avail. On Sunday morning, I watched the family again, but without success: All of them spent the whole time looking for food.

The matter gave me no peace, and finally—it was again just a gut feeling—I decided to visit the cranes again in the late afternoon. To my disappointment, the family was not in their usual spot. I waited. Dusk began to fall, and suddenly three cranes flew in, two older birds in front and behind, and a young one in the middle, clearly recognizable by its brownish head.

I was unsure: Was this the pair from Klein Hansdorf, whose young had been fledged for 2 months? I doubted this, because the cranes did not land, but rather flew in wide arcs over the moor, over the territory of the family I actually wanted to observe. But for the Hansdorf family, it would have been the totally wrong territory, a few kilometers away from theirs.

I could not explain the situation, but I was surprised that there were no territorial defense calls from the territorial pair against the intruders. Therefore, I concluded that the flying family I was watching was the one that I had originally wanted to observe. So, I had now seen my longed-for first flight—perhaps it was actually the second or the third flight of the day, but, in any case, this was definitely "first flight day". Yesterday and this morning, the youngster had been unable to fly; this evening, it was already flying figure eights and circles, and even made an excursion beyond the territory's boundary.

Then, the cranes landed, which is when they really began to captivate me. Because, to my surprise—it was quite dusky after all—they started looking for food. And the juvenile begged its parents to be fed. But father and mother remained strict and did not feed their offspring: They led it to nutritious places, pecked around in the grass, turned over sods, but only showed what there was to find—the young one had to eat by itself. Eventually, it did, but obviously reluctantly.

It seemed very tired and hungry. All three ate for an unusually long time. The young one was also obviously thirsty. One after the other, the three got into a ditch and drank by dipping their beaks, then bending their necks sharply down and back—with the beak still in the water—and lifting their heads so that the water could run down their throats.

Finished, dinner was over—but the spectacle continued! Now, instead of preening their feathers, or walking to the roost site, or flying there, the adult birds began to dance. It was the middle of September, and the parent birds were dancing as if in the middle of the best spring courtship! They were

Fig. 3 In Arjuzanx, southwest France, is a big wintering range for Common Cranes. Here, two juveniles (recognisable by their brownish heads, about 6–7 months old) in their first year following a parent crane: These had got enough strength to endure their first migration from northern Europe. (© Bernhard Wessling)

jumping up, flapping their wings, dancing around each other, calling and trumpeting.

The juvenile was perplexed. He had never seen anything like it. He just stood there and watched. His parents danced long and hard; they did a real dance of joy. And finally, the offspring felt animated to imitate. Delightfully awkward, he jumped up, flapped his wings, wanted to do it just like his parents, but couldn't get it right. A Disney animated film could not have been funnier; rarely have I laughed so loudly at an animal event.

It took a long time before the family calmed down and slowly—it was almost dark now, I could hardly see them—retreated to the sleeping place. I very thoughtfully walked the way back to the car.

What I had seen had once again been unique: A young crane had flown for the first time, the family had made practice rounds, the juvenile in the middle, the parents often looking after it. The flights succeeded without a crash, the landing was almost perfect. Then, there was eating and drinking. Afterwards, an irrepressible joy broke out! I can't interpret it any other way.

I don't know if crane parents are always so happy when their juveniles fledge. I don't know if any other crane observer has ever made a similar observation. I have never read anything about it, and even after I reported it at conferences, I did not hear or read about anyone who had experienced something similar.

I am convinced: Cranes are much more aware of themselves than we want to give them credit for. They "know" that their young bird must fledge. This particular pair may have known when they bred that it was late in the year and that the second breeding attempt was a risk, because it would be mid-September before the first flight could be possible. From there, there were, at most, 2 months until their departure for Spain, during which time the young had to acquire the fitness for a long flight.

The crane parents were simply delighted that another milestone had been reached along the risky path! And I with them (Fig. 3).

In Search of the Cranes' Language: They Call and Thus Tell Us About Their Lives

Over the years and with the growing number of crane pairs in the nature reserves that we managed, our observations became more and more confusing. Ever more new questions arose. To name just a few: Is it really the "Sly Dogs" that breed sometimes in the west, sometimes in the east of their large territory? And if it is, why have they now lost their conspicuous shyness, at least outside of the breeding territory? Are the "Pioneers" still the original "Pioneers" or are these now different individuals? Do crane pairs really stay together "forever", or at least for a few years? Are the "Seekers" in the "Green Brook" this year the same birds that investigated the territory last year? And where have the "Townies" gone? They couldn't have been that old after only 4 or 5 years of breeding. Did they just disappear, or did they take over a completely different territory after a year off from breeding? Did they become displaced, or did they switch voluntarily? Has there perhaps been a change in partners, and thus a change of location?

Questions that can only be answered if the individuals can be clearly identified and recognized. With birds, this is usually achieved by banding them. But we could not and did not want to ring the cranes here. The Brook is a much-visited nature reserve, and it is against my view of nature conservation to pursue the cranes that we ourselves are protecting with a hunting team so as to catch and ring them. Such an action would also have contradicted the understanding of nature that we wanted to convey to the people of Hamburg and Schleswig-Holstein: to respect nature and not use it recklessly. Banding has another limitation: Usually, you only catch young cranes before they fledge. That means that we would not have banded crane pairs, but very young cranes who may or may not come back to our nature preserve, and if they return and start breeding, they will have an unbanded partner who could have come from anywhere else. But catching adult cranes, especially pairs, is much more complicated and only possible during moulting. How to attach rings to a complete population of *adult pairs* in a certain area like ours? Therefore, collecting information as I wanted it was not possible by means of banding.

The banding of birds for scientific, i.e., human, purposes of gaining knowledge is also a kind of utilisation. In addition, banding operations eventually kill individual birds, or injure them so severely that they cannot survive in the wild afterwards, i.e., they have to spend the rest of their lives in captivity—losses that have always seemed irresponsible to me. This is not to say that I am against banding in general. In areas where there are numerous cranes and very few people, ringing is fine. It is useful and helps us to understand the migratory behaviour and regional flexibility of cranes.

The same applies to the attachment of radio transmitters. Undoubtedly, this is an elegant modern method, with the help of which one can investigate the occupancy of territories, flight activities and even migration routes. However, for the reasons mentioned above, this would not have been appropriate in our area either, especially as there would have been no one willing or able to pay for, carry out and evaluate such an operation.

The stress to which the cranes are exposed here is incalculable. And because I thought the birds were capable of learning, I wanted to convey to them that people would stay on the paths, wouldn't get too close to them and don't want to do anything bad to them. I didn't want all the cranes in our Brook, an area with a high human visitor density, to be afraid of humans year after year, because they knew from experience that, at any moment, some humans could jump out of the bushes and encircle, catch and ring a young crane that had not yet fledged.

In 1996, I began to think about the possibility of distinguishing and identifying cranes by their voices. I had the impression that I could tell the calls of different crane pairs apart by sound and pitch. My good musical ear, honed through playing the violin in my youth and regularly playing the piano, helped me in this. My friends, however, thought that the differences I heard were a product of my imagination.

In 1997, I heard a particularly characteristic call in spring. It came from the male of a new young pair that did not yet have a territory. Unlike all other cranes, this male had a very raspy voice. I therefore named the pair the "Rattlers". In the meantime, I got the idea to try recognizing individual cranes by recording their calls. I first did some research, because I could not imagine that I was the first to come up with this idea. And I wasn't: Ten years earlier, Eberhard Henne had already observed and recorded several crane pairs in half a dozen territories in the Schorfheide Biosphere Reserve (Brandenburg) over a period of several years. (By the way, Henne, who had spent some time as Environmental Secretary in the state of Brandenburg, headed this Biosphere Reserve for many years.) He intended to evaluate the crane calls with the help of spectrograms. A spectrogram is the pictorial representation of an acoustic structure over time.

When Eberhard Henne and I met at a crane conference, he already had a considerable archive of crane call recordings. Unfortunately, he had not yet succeeded in recruiting bioacousticians who had suitable methods for sonagraphic evaluation, so that he did not yet know whether it would be possible to reliably distinguish cranes on the basis of their voice and to recognize them in subsequent years.

For his recordings, Henne used two old-fashioned ("made in the GDR") and heavy magnetic tape recorders. In mid-March, he drove his car to a few easily accessible crane territories, where he wanted to find out whether they were still the same as the year before. There, he played unison calls from the first device early in the morning—the cranes felt provoked by this feigned territorial violation and called back. Eberhard recorded these calls with the second device. Sometimes, he could not stop the first device and start the second one fast enough to record, so quickly did the respective territorial pair answer.

Compared to us in the Brook, Eberhard had an easy job. He was in charge of a huge area of over 130,000 ha, where 180 pairs were breeding at that time (there are even more today), and he chose those territories for his call recordings where he could get close to them by car with little effort. We talked about his experiences and my goals at the crane conference and immediately decided to work together.

At the same time, I asked a biology student from Hesse, who was doing a crane watching week with us, to visit her university library and research what methods could be found there for analyzing bird calls. It was possible that something useful already existed that I could easily buy or copy. But she found nothing, not even reports of how other ornithologists had studied bird populations of any species using vocal analysis and observed them over time.

I finally found a bit of software ("avisoft") that was based on analyzing the melody of bird calls. Unfortunately, it proved to be unsuitable for my purposes. On the one hand, the unison call of the cranes was difficult to analyse with its help, because the melody is "sung" (more accurately: "trumpeted") by two birds. Secondly, in the wild, it is often not possible to record the cranes' calls from the first second on, which is what the software required.

So, I had to develop another technique. To prepare this, I bought a very powerful directional microphone, connected to a digital recorder and mini-disc. All the equipment was easy to transport. I could rely on the cranes call-ing at some point in different life situations, so I would have plenty of opportunity to record their calls. I just had to react quickly when a call started, and the remote control of the minidisc allowed for this. Early morning unison calls from inside of their territory, when dawn had barely begun, were particu-larly fruitful. At this time of day, I knew I could hear the territory owners in *their* territory. Often, the responses of neighbour pairs could be heard and recorded as well.

Later, in other areas (in the nearby Nienwohld Moor, in Mecklenburg, Brandenburg, in Japan, Korea and the USA), I played back suitable unison calls with a small transportable CD player, causing the sound to carry over distances with the aid of a big megaphone—and thus I provoked the unison call of territorial pairs in the vicinity of where I was standing.

My older son Bengt and the aforementioned biology student helped me with the first recordings. Since we did not have as many pairs to choose from as Eberhard Henne, we had to record the pairs on their grounds, no matter how inaccessible they were. We ourselves ensured, through our conservation measures, that the water level was always good, at least in some territories, and so we could record calls in most territories only from a great distance. This made the recordings difficult, and particularly tedious. We often had to wait a long time, especially in the morning. Since we were doing our crane moni-toring and observing anyway, the time required didn't matter, but the proce-dure and preparation were more complicated. (Just plug all of the cables together at half past four in the morning and press all of the correct buttons on the electronic devices, all in the dark! If you have not left any cable or plug in the base three times in a row and if you have not forgotten to recharge the batteries the night before, you have passed the first examination level for access to the Duvenstedt Crane Bioacoustics Society.)

In each case, there were between 150 and 500 meters, sometimes even a kilometer or more, separating us from the cranes. Over this distance, we had to hope to be in the right place at the right time and ready to record.

Fortunately, the cranes helped us (Fig. 1). After all, in 1998, we had six ter-ritorial pairs on the approximately 10 km^2 of the nature reserve and the nearby surroundings, the "Darned Seventh" was hanging around, and groups of juve-nile cranes also came by and spent the night in the central area. Three pairs lived in close proximity to each other, at a distance of only 100–150 meters, and the next one had its territory about 500 meters away. Two further pairs

Fig. 1 A Eurasian Crane pair is unison calling while their chicks don't understand what's happening with the parents. (© Carsten Linde)

had only about 200 meters separating them, but were about a kilometre "as the crow flies" away from the others.

In the early morning, almost all of the pairs called each other. If everyone called and there was no wind, it was theoretically possible to record all pairs at around half past five/six in the morning (summer time) at the beginning of April (1 hour earlier in mid-March). In fact, however, we needed 4 weeks and 140 recordings to get recordings of sufficient quality from all pairs and to be sure that the recordings and voice analysis were reproducible.

We additionally recorded during the day, depending on where we were, and observed cranes. Sometimes, strange situations occurred. For example, I was cycling looking for the seventh pair in the south of the area with the recording equipment on hold, when Bengt informed me by radio: "If you ever want to record the 'Sly Dogs' from very close, that is, when they're only 50 meters away from you, you'll have to come to the 'Corner Pasture'." This was a spot where this pair often stood relatively close to a fence that year, pecking around in what must have been a very nutritious mud pool. Needless to say, I very much wanted to have the "Sly Dogs" toot right onto my minidisc. But first, I had to drive through the whole Brook, and by the time I arrived at the "Corner Pasture", the cranes could be long gone.

So, I hurried. Bengt had departed, so as to give them no reason to leave. I arrived 15 minutes after this, after a quick "Tour de Brook", and surprisingly, the "Sly Dogs" were still standing there, quietly foraging. I got off the bike outside of their sight, switched on the equipment and moved slowly along the path towards the fence.

And just as they had a few minutes earlier with Bengt, they started to give out a unison call. I guess that meant that I was unwanted and should leave. I didn't get the impression that they were particularly afraid, because they continued starting and stopping the call while maintaining visual contact with me. After a minute, when the desired recordings were on the minidisc, I did them the favor and disappeared again.

Also helpful for recording calls are "airspace violators". If you know that a pair is in their territory, you only have to wait for an unfamiliar crane (or another pair) to fly over. You can then "harvest" a nice unison call from the territory owners. To do this, however, you have to anticipate, based on previous observations, when and where cranes that don't belong to the territory might show up.

Once, when the male of the "Rattlers" chased off six foreign cranes, I got to hear some interesting flight calls. Four of the strangers landed in a forest pond directly behind my location on a small hiking trail, two of them calling afterwards. This gave me a new call pattern that I could use to recognize these cranes if they showed up at our location again.

Only through the acoustic documentation did what we had suspected based on our visual observations, but could not prove for a long time, become 100% clear: That we had a seventh pair in our area in 1998, that a single crane flying around was actually single (maybe the widower "Romeo") and that there were also six juveniles roaming around.

Because I listened to the numerous recordings over and over again during my analysis work, it soon became possible for me to clearly determine at least four of the seven pairs just by listening. If I had the direction as an additional hint, I could tell all six breeding pairs apart—something that had been unthinkable before.

I was able to record many other calls besides the unison call, because I was interested not only in the individual characterization, but also in what distinguishes the other calls (flight calls, contact, warning and search calls) from the unison call. Perhaps this will one day allow us to draw conclusions about how cranes communicate.

Soon, thanks to my high-performance microphone, I learned that cranes exchange very soft vocalisations before they jointly take off, because, using my laptop, I could later hear what I had not perceived with my ears during the recording. So, cranes communicate not only by body language, but also vocally. This "take-off coordination vocalisation"[1] is different from the very quiet "cooing"[2] or "rolling"[3] contact sounds, which I already knew about from the close proximity encounter with the "Sly Dogs" at the hunter's blind, and that I was soon also able to record, when I was allowed to listen to them from a close distance while well hidden.

Very quickly, I realized that, if many calls of many pairs were to be compared, the spectrograms could not easily be used to identify individuals. I needed an additional analysis method. So, we searched all of the literature again. The student that I mentioned earlier searched the general biological literature—there just had to be papers by other researchers! But there weren't.

I myself searched the literature for mentions of bioacoustics in regard to cranes, a task in which the ICF, with its well sorted library, was a great help. However, except for a few papers, nothing could be found there that could have helped me to develop an evaluation method.

We unearthed a single publication that attempted individual characterization of birds using bioacoustic methods: woodcocks from widely separated areas were studied. But birds in neighbouring territories were not studied in this research, let alone over several years. So this work didn't get me anywhere either.

In the case of cranes, the "melody" and rhythm of the call is quite similar, but has very subtle characteristics that vary from individual to individual. My initial hypothesis was that the structure of the sound-producing organ, the syrinx (the "vocal head"), which is located on the trachea, determines the frequencies with which an individual can call. The vocal head has membranes that can be modulated by muscles as they vibrate during the forceful expulsion of air (when the bird wants to call).

These oscillations should, I thought, be different and characteristic for the individual cranes and hopefully remain relatively stable over years. To determine this, my analysis method had to be sufficiently sensitive. After testing several computer programs, I decided to use two of them to analyze the calls.

---

[1] Take-off coordination vocalisation of grey cranes: https://soundcloud.com/bernhard-wessling-225998183/14preflight/s-xeAS2

[2] Contact sounds of Eurasian Cranes: https://soundcloud.com/bernhard-wessling-225998183/22-contact/s-vzym0

[3] cf. footnote 2 in chapter "Arrival in the Brook After Returning from Wintering Grounds: Alone or in Groups?".

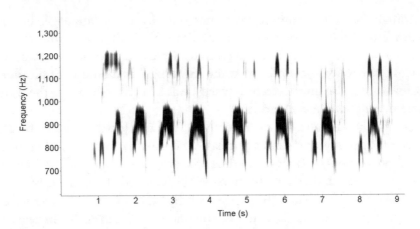

Graph 1 Spectrogram example, in this case between 1000 and 1200 Hz, the female's calls can be seen, the male's are below 1000 Hz

One, "SoundForge", results in the so-called spectrogram, which displays the frequencies (i.e., the pitch) as a graph over time; the volume is expressed through color intensity.[4] This allows you to see the progression of calls.

I developed the second method with a software application engineer based on the platform "mathematica"; we first created a 3D set of data (Graph 2), from which a so-called "power spectrum" could be generated using two complicated calculation steps (which we programmed).[5] I call such a spectrum the "acoustic fingerprint" or "voiceprint"; it shows the loudness per frequency, summed up over the entire call period. To compare crane calls with each other, I compare power spectra (Graphs 3–6).[6]

Before any analysis, I removed unwanted sounds, such as an interrupting cuckoo or oriole, using SoundForge. Tits, sparrows and similar small birds do not interfere, because I only look at the spectra between 600 and 1200 Hz (the frequency range in which most Common cranes call), and these smaller birds sing or chirp at higher frequencies. The overtones (harmonics) of the human voice are within the same frequency range as the cranes' fundamental frequencies, so I almost always went to the Brook alone, without any company, for the call recordings. This was necessary because, although I decreed "absolute silence", a companion would inevitably speak in between calls, often with a spontaneous outburst such as, "Do you hear the cranes

---

[4] An spectrogram example can be found at https://www.bernhard-wessling.com/sonagram-example (Thanks to Dr. Frommolt, Museum für Naturkunde and head of the Animal Voice Archive, Berlin, for making me better spectrograms than I could make with SoundForge at the time; the colourful section at the top right of the picture is from a SoundForge spectrogram).

[5] Example of a power spectrum: https://www.bernhard-wessling.com/powerspektrum

[6] Here, you can see one example each of calls from the same pair, from two different pairs and from a change of partner: https://www.bernhard-wessling.com/powerspektren-vergleichen

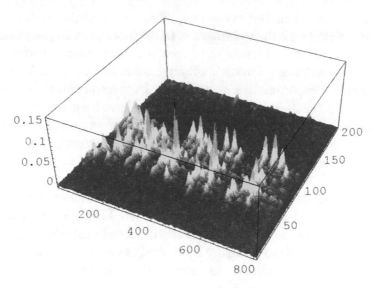

Graph 2 Illustration of the three-dimensional set of call recordings data

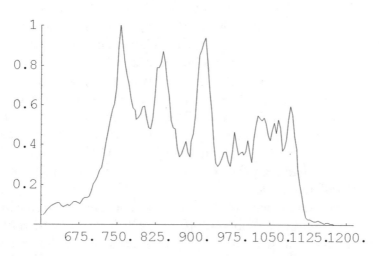

Graph 3 The power spectrum created therefrom by summarizing all intensity data in narrow 5 Hz frequency windows, thus eliminating the time axis

calling, too?" Yes, I hear them, I am recording them, and now that recording is ruined.

The spectrogram reveals to us the shape, the melody of the call, the "power spectrum" ("acoustic fingerprint" or "voiceprint") the characteristic frequencies. Generally, the "power spectrum" is enough for me to identify a particular individual crane or pair, but sometimes, there are uncertainties that can then be clearly resolved with the spectrogram. In the more than 10 years that I have been using bioacoustics, I have always done the comparisons "manually" (i.e., visually, by comparing printouts), although all data were digitally available. An automatic comparison of a new call, i.e., its spectrogram or "acoustic fingerprint", with the digital library of calls would have been very desirable, but is still not reliably possible; at least, I have not found any suitable software for this. Right now, I am once again working on this question, together with a doctorate student, so far with no result.

While, at the beginning of my recordings, I first made sure that the unison calls of cranes from certain territories remained uniformly characteristic over a season, it was still open as to whether this would also be the case over a period of years. I decided to listen to Eberhard Henne's recordings and analyse them: Would I be able to recognize individual cranes in them, or even determine if there were different cranes in one of the established territories than had been there in the previous year?

Eberhard gave me his "treasure": numerous magnetic tape cassettes in a plastic bag and one of his old tape recorders. I set up playback and recording equipment in my study, and, with Bengt's help, I transferred the recordings to my minidisc over the course of several days. Ultimately, I had about 200 digital files from 7 years with recordings for analysis.

The analysis clearly proved that cranes can be recognized by their voice, even over several years. Over time, a change in the voice was detectable, a kind of "maturation" as occurs in humans. The frequency pattern of the "acoustic fingerprint" is so characteristic that pairs and individuals can be reliably distinguished. My evaluation method brought to light plenty of surprising results. After all, I had been able to evaluate about 40 "crane pair years" from a total of 13 different pairs.

Eberhard had warned me: "In 1996, two pairs of cranes were breeding very close to each other in the 'Bibberstipp' area, which surprised us very much! They even bred within sight of each other!" After analyzing his tape recordings, I was able to tell him more about his cranes than he himself had known: It was not only that there were two different pairs in 1996; in addition, the second pair had already been in the same territory in 1994 and 1995! In 1997, the older pair was either not there or had not called during the recordings, while in 1998, both pairs (with the unromantic names of No. 6 and 7, respectively) were present again.

What's more, such a thing seemed to happen more often than we would have expected. In 1995, pair no. 13 joined pair no. 10 AB in the "Fuchsbruch" area. In "Zabelseebruch", in 1994, pair no. 4 was found next to pair no. 3, as were pair no. 2 next to pair no. 1 in "Zabelsee" in 1996. So, from 1994 to 1996, there was always at least one "double" occupied crane nesting ground somewhere in the investigated area of the Schorfheide.

Something similar had happened in Duvenstedt Brook, so it was not too big of a surprise for me: When the "Foxes" bred for the first time, and only about 200 meters away from the "Pioneers", this remained hidden to most crane watchers and observers; there were even debates about whether they were really two different pairs. The same thing happened when the "Rattlers" first appeared in 1997. Most of my fellow crane watchers doubted that it was another (the third!) pair in close proximity, holding the view that one of the two pairs, the "Pioneers" or the "Foxes", was sometimes seen here and sometimes there. Before the days of bioacoustics, the truth could only be ascertained by visual observation, when several persons armed with radios posted themselves in three places at the same time—this was seldom a viable option and was very difficult to organize. Through multiple observations, one had to prove that the individual pairs flew out of certain roosts in the morning or flew in in the evening; only then could one be sure that there were really two or three pairs breeding in close proximity. Thanks to the voice analysis, the recording of morning unison calls was now sufficient to tell how many crane pairs there really were and where they lived (recordings needed to be made several times during the season to prove that the pairs actually occupied the territory there).

In Eberhard Henne's crane territories, however, pairs sometimes nested less than 50 meters apart; I had not witnessed anything that close in our area. While some "double occupations" in the Schorfheide lasted only one season, pairs no. 6 and 7 apparently got used to each other as neighbours. During the first evaluation, when I only had the recordings up to 1997, I still thought: "Ah, that's what the apparent continuity of cranes in territories is really about: For some years, only pair no. 7 breeds, then, for a few years, two pairs breed together in close proximity, which often cannot be determined. Then, the younger pair, in this case, No. 6, takes over the territory." But in 1998, pair No. 7 was also back on site.

In general, the cranes in Eberhard's territories were faithful to their partners and their territories. In the 40 "crane pair years" evaluated, territories have been reoccupied 34 times, 29 of which were by the same pair (representing 85%). The most stable pair of all was No. 7, which was found in the "Bibberstipp" from 1991 to 1998 (apart from one year in which it perhaps took a break from breeding).

But there were exceptions. In the "Fuchsbruch", we had pair no. 10 (male 10A, female 10B) in 1991. By 1994 (unfortunately, there were no recordings of them in 1992 and 1993), 10A had separated from his partner B and had taken up with the female 10C. These two were still breeding in 1995, in a neighbourhood with another pair. Regrettably, we were again missing the years 1996 and 1997, but the two had disappeared in 1998, at the same time that a new pair (no. 13) had appeared.

In the period from 1988 to 1998, for which, however, some years of recordings are missing, I was able to identify four different pairs in a certain territory through bioacoustics, and even to recognize the change of partners in one of them. Purely visual observation would have suggested the normal continuity of a pair in the territory.

In 1994, we find pair No. 9, male 9D, female 9B. In 1995, it is replaced by pair No. 8, male 8A, female 8C. The next year, however, we find a newly combined pair of male 8A and female 9B—did 8A choose 9B or did female 9B reconquer her territory, "take over" the male living there and drive away her rival? This pair then remained stable: In 1997 and 1998, we found them again in the same territory. In total, we had three "divorces" (that is, about 10%).

This behaviour may surprise us, especially the established crane experts, as we have always considered cranes to be the epitome of faithfulness when it comes to partners and territory. But doesn't such behaviour make it clear that we are dealing here with individuals capable of making decisions, with individual characters and different personal experiences; and not just with instinct-driven breeding machines? Or do we find the behaviour so strange because it is so similar to our own and runs counter to traditional human moral concepts?

Also interesting are pairs that we hear only once. In "Zabelsee" and "Zabelseebruch", pairs no. 2 and no. 4 were each found in only one year. In "Fuchsbruch", pair No. 13 appeared only once (or, it was recorded only once). Perhaps they were not tolerated, perhaps they found a new, better territory for themselves. What we learn from this is: Cranes actively seek out territories. It happens that, for some time, they exist in a territory that is already occupied, and are temporarily tolerated there by the territorial pair. But for reasons we do not (yet) know, they move on.

But we also need to consider that Eberhard recorded his crane pairs' calls usually only once per season. I preferred to obtain many calls per pair in each season, at least ten to twenty, but always hoped for more over the duration of the season.

In the breeding season of the second year of my call recordings—it was the year 1999—I was able several times to be ready as early as four o'clock in the morning to record unison calls in the Brook, before I drove to my company at eight o'clock, or I walked through the Brook in the evenings and on the weekends with a microphone and remote control in my hands and the recording devices in my backpack. I would analyse the calls I recorded at home on the computer whenever I had time.

To my surprise, I discovered the following: The "Pioneers" were still a pair, although I would have bet that they would break up after the double brood loss in 1998. Without the clear results from bioacoustics, I probably would have thought there had been a mate change in the meantime, because the habits of this crane pair had changed. After all, it had been the relative constancy of the behaviour of all pairs in the previous years that had caused me to believe that I could recognise certain pairs or individuals. The bioacoustic analysis, however, *proved* to me that it was the same pair, although it almost never stayed on its former pastures in the north of the Brook, but now preferred another meadow and had even been seen (i.e, heard) intensively examining a damp woodland in the northern Brook (probably as a potential new territory, from where I could record additional calls). Eventually, however, it bred in the same place as in all previous years, just much later. Also, breeding was unsuccessful that year.

All other pairs also changed their habits, some quite drastically. Without bioacoustic analysis, for which I recorded calls on many days at all times of the day expressed from various locations, I would certainly have drawn quite a number of incorrect conclusions, or at least would have been very uncertain, because almost all crane pairs had chosen new feeding meadows and/or moved their nesting sites. All of this and much more would not have been detected had I not analyzed hundreds of call recordings: The "Rattlers" moved their nest 300 m to the west, started to breed, but then aborted this breeding attempt; a little later, they started a successful second breeding—and this in the middle of the territory of the "Planners". I was only able to determine that it was, in fact, this pair that had recently bred there "on foreign soil" with the help of bioacoustics. For the "Rattlers", breeding at this place was only possible because the "Planners", who had bred in their previous territory, did not guide their young out onto "their" previous meadow, the "Long Pasture", but instead to a sparse bog birch forest east of the main path.

The "Foxes" no longer flew to the "Hill View" to feed, but flew only 150 m south from their breeding place. The "Sly Dogs" had also changed their habits: They almost never appeared in the feeding meadows of previous years, in the northeast of the Brook, not even in the "Corner Pasture" where they had trumpeted at me the year before; without the bioacoustic analysis, I would certainly have doubted whether they were still the same pair.

The "Seekers" apparently moved into their territory in the middle of the "Green Brook" quite late, when the season was almost over. Before that, they seemed to explore other territories, for example, the "Small Moor"—all of this

I would have suspected without bioacoustics. However, vocal analysis showed me that the calls that I had recorded at the beginning of the season in formerly unused territories, including the "Little Moor" and south of the Brook, belonged to two new, previously unknown crane pairs, which apparently stayed in the Brook for only about 3 weeks each, found that they were not tolerated in the territories, and moved on. In fact, the "Seekers" were not there at all, although they had already bred the previous year, albeit unsuccessfully.

Instead, at the end of April, "the Darned Seventh Pair" had taken over the former territory of the "Seekers", i.e., the pair that had already been prowling around in the south of the area in 1998 and that (according to my guess) had been constantly harassed by the widower "Romeo". Now, the pair was a year older, but by no means called in unison, instead issuing single calls that were not yet developed. It was not until the end of May that I heard the pair calling in unison for the first time—a new pair had taken over an abandoned territory.

Of the six most recent years' territorial and breeding pairs, five had returned; the sixth territorial pair was identical to a pair that had been on the move in the Brook the previous year as "prowlers" without a territory. Even then, I had only noticed their presence because I had bioacoustically evaluated the calls of all cranes in the brook, not only territory owners. In addition, at least two new pairs had arrived in the area and were either looking at territories or were involved in territorial disputes. In addition, there were groups of up to ten juveniles that did not make any acoustic noise. They often stayed in the brook only overnight.

What would have been a fuzzy, puzzling and confusing game for someone who only observed the cranes with binoculars was an open book with the help of bioacoustic analysis. We now knew that territorial occupation and selection, as well as territorial and mate fidelity, were much more complicated than we had assumed. Cranes are individuals that actively engage with their mates, neighbours, competitors, and environment. They observe, they decide, and they develop and change their habits.

Despite years of intensive observation of the cranes in the Brook, surprises were still possible in retrospect. During the evaluation of the first recording years, I approached an acquaintance who, for years, had been taking videos in the Brook, mostly of red stag and fallow deer. I asked him if by chance he might have recorded some nice sequences of cranes with unison calls and if he could let me have these scenes. To my delight, I received several videos. When I watched them for the first time, I listened in surprise: There were two cranes calling with warning calls, at first sounding very much like the "Darned Seventh Pair" that had appeared in the Brook as recently as 1998. But that couldn't be, because these video recordings had been made in 1996. What's more, the scene concluded with a unison call! So, I created a digitized sound version and analyzed it. The result: The warning call was melodically similar to that of the "Seventh Pair", but that was where the similarity ended: The pitch of both the male and female was almost half a tone lower than that of the "Seventh Pair". And the acoustic fingerprint of the calls looked completely different. Who were those cranes on the video?

The unison call solved the mystery: It was clearly the "Seekers" who had occupied a territory for the first time in 1998! So, they had not only visited the Brook in 1997 (as I believed I had observed), but had also already done so one year earlier (1996) while investigating territories, without us having concluded as such from visual observations. The video scene does not show the territory that they occupied two years later, but rather another one. Watching it, I realized that the "mismatched pair", as we had called them at the time, were the "Seekers." And the "Pioneers" were already acoustically recognizable on a video from 1996.

Systematic bioacoustic studies had been virtually non-existent in the crane world prior to this, although I noted that some crane researchers and conservationists had considered it or had already tried it. The Japanese crane researcher Professor Hiroyuki Masatomi told me, when we met in Japan, that years earlier, he had given up on the idea of recognizing cranes individually by voice recognition, but now he was happy that, due to my new method, his old dream had come true.

George Archibald was one of the first to deal with the bioacoustic study of cranes in his doctoral thesis. However, his goal at that time was different. He wanted to verify the relationship of the cranes to species and to each other by analyzing the *form* of the unison calls. To do this, he recorded calls from all the crane species in the world and actually found some inconsistencies. His new assignment based on the unison calls was later confirmed by genetic analyses. Individual characterization was neither intended by nor possible for him at the time (decades ago, preparing spectrograms was based on mechanical devices that produced rather crude spectrograms). But George was already convinced, even then, that each crane had its own distinctive voice, just as I had heard after so many years of listening to the calls of the Brook's established and newly arriving pairs. Therefore, George was the first crane expert to accept my new method without hesitation and make it known.

A second study, devoted to the Sandhill Crane, had the aim of elaborating individual and regional differences in calls. The study had been conducted in such a way that made a mockery of any idea of nature conservation. The researcher had borrowed a dog that was trained to detect crane nests. The dog's owner had taught the animal to do this in order to conduct a regular census of crane broods in his vicinity. The researcher walked this dog to the nest, thereby startling the crane that was breeding and causing it to utter the so-called "guard call". However, she was able to detect individual differences. But as a method of data collection, I ruled out this approach, because it is based on disturbing the crane at the nest. Moreover, it does not allow for the detection of pair bonding (only one bird is at the nest, the other is somewhere else, and only one bird calls, and you don't even know if it is a male or a female).

A third study aimed to determine the sex of Whooping Cranes. The "guard call" was also used for this purpose. But this method cannot be used to corroborate behavioral observations. I later learned that there was another collection of Whooping Crane "guard calls" from the wintering area in the Aransas National Wildlife Refuge in the Gulf of Mexico. This work had not been published, but I obtained the recordings for analysis. Regrettably, they proved to be completely useless.

All of these unsystematic and rudimentary attempts were not suitable to identify and recognize cranes or crane pairs individually, let alone to observe a larger area or a crane population over a longer period of time.

With my method, it suddenly became possible, on the one hand, to create the history of a crane population in a certain area (here in the Duvenstedt and Hansdorf Brook) over several years and, on the other hand, to write down individual biographies of some pairs without ever having banded the cranes, i.e., without ever having disturbed them.[7] Such a study is not possible with only a few recordings of individual calls before and during the breeding season. Rather, I wanted to get hold of 2–20 recordings for each pair in order to be sure that their frequency spectrum was characteristic and unambiguous and that a pair could be assigned to a territory. In this way, I also made sure that I could "listen" to significant parts of the season dynamics for several pairs in our area. With only one or two recordings per season, you will get only a short flash of the real events.

As the number of territorial pairs in the Brook was constantly increasing, this meant that there were more than 200 call files per year that could be analysed (I deleted many more call files for various reasons or simply found that I could not use them). On each day that I was out with the microphone and recorder, I received between zero and, perhaps, ten evaluable calls. Now you can calculate how many times I went out "hunting calls" over the course of

---

[7] Overview table: www.bernhard-wessling.com/table-overview-10years-brook-sonagraphy, plus isolated analyses of video sequences from 1996/97.

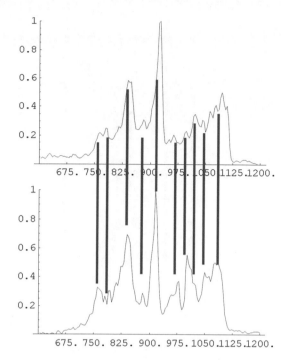

Graph 4 Comparison of two power spectra generated from two different calls of the same crane pair in subsequent years

3–4 months. Evaluation is also time-consuming and requires patience, diligence and a high level of frustration tolerance (Graphs 4, 5, and 6).

In 10 years of bioacoustic monitoring (plus analysis of video sequences from 1996/97), I got to know a total of 25 crane pairs. In the last years of my study, we had twelve territorial pairs. Because many of them are not faithful to each other for life after all, there were 61 instead of only 50 different crane individuals. I was able to document 112 "crane pair years". 18 pairs were in the Brook for at least 2 years, 14 pairs for at least 3 years. The latter group alone represents 80 crane-pair-years, averaging almost six crane-pair-years per pair.

Only two crane individuals (M4, the male of the "Foxes", and F2, the female of the "Planners") were present in each of the ten observation years (although not always with the same partner). I was able to detect pair No. 5 in 9 of the 10 years, male and female always being together; it is the only pair in my study that was inseparable for so long. Pair No. 4 was also heard and seen in 9 years, but in one of the years, M4 appeared with a new female, and I thought, "Well, F4 must have died." But that wasn't the case, because the following year, M4 and F4 were a pair again. The "Planner" female was with the

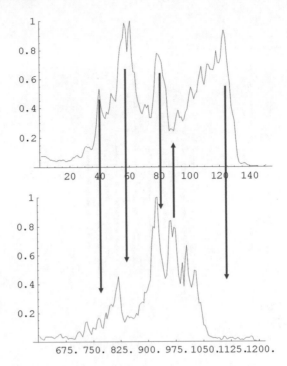

Graph 5 Two power spectra of calls from different pairs showing the peaks and valleys at completely different frequencies

"Planner" male M2 for 5 years, then she appeared with a new male, M19, and shifted territory by only about 300 m. So, she was the one who determined where the pair settled (this is unusual; usually, the males are the "territory owners"). The relationship outlived a short "affair" by M19 with a newly appearing female F30 in another territory in the following year, the same year that M19 and F2 bred together again.

Six of the 14 pairs (i.e., 42.9%) that had territories for at least 3 years were faithful to each other for those 3 years or more (as far as I can deduce from unison calls, that is; I do not, of course, know what happened in the silence and darkness of unobservable areas). Of the eight other pairs with a minimum of 3 years' presence in the Brook, with 57.1%, the clear majority had changed mates that could be heard and analyzed. M6 (the "Rattler") was present in 9 of the 10 years and had four different females during that time. At the same time, he was the most successful crane during this observation period, with eight fledged offspring.

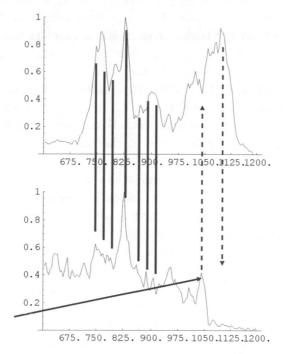

Graph 6   These two power spectra show a different pattern in the range of the female frequencies: Male 6 (formerly paired with female 6) then together with female 11

Only two pairs always occupied the same territory during the whole observation period; all other pairs changed territories. Thus, we see that some things are different from what the crane experts say and write about them.[8]

In the 10 years from 1998 to 2007 that I listened to the cranes of Duvenstedt Brook, they became more and more familiar to me. Some pairs I thought I knew well, others not so well, still others were strange to me. In using terms like "knew" and "familiar", I mean that: I knew which cranes they were; I knew where they lived; I knew when they had taken possession of their territory in the Brook; and I could observe and document how many fledglings they had raised. With some of them, I had experiences of the kind that I have described in the earlier chapters.

---

[8] Even "Kranichschutz Deutschland" (the organisation responsible for practically all crane protection and research activities in Germany) still writes on its website https://www.kraniche.de/de/faqs-haeufige-fragen.html to the present day (last visited on 1.4.2019, translation by the author): "4. Are cranes faithful to each other? Yes, cranes live in 'permanent marriage'. However, banding has shown that there are also mate changes. A new mating occurs above all in the case of illness or loss of the partner, as well as when the male is unable to defend the breeding territory." My research showed that this is not correct; sure, these are also reasons for a crane to look for and find a new mate, but I have shown, in many cases, that the former previous partner is alive and also looking for a new partner or had already found one.

Perhaps I sometimes imagine too much, for example, that I know a lot about the motives of the birds, or that, when they are not calling at the moment, I still "know" who is striding across the meadow or through the birch grove in front of me and what they may soon be up to. The recordings and analyses have provided me with a lot of information, and I would like to report on some pairs in more detail below. In doing so, I will endeavour to keep the facts on the one hand and my conjectures on the other clearly separate. The reader will recognize some of the crane pairs from the previous chapters.

The **"Pioneers"** were certainly present in the Brook in 1996 at the latest, occupying the central territory ("Great Moor"). We know this because, in that year, an acquaintance happened to have made a video recording of a unison call, which I analysed in 1998. In the early years of my "friendly eavesdropping" on the Brook, this pair was a regular presence. The "Pioneers" were at least 4 years old in 1996, but probably older. I also called them "pair no. 1,1" in the meantime. I had changed to numerical designations because I had noticed again and again that pairs changed their territories. Therefore, the place names originally used (like "from Great Moor", i.e. designated "GM") no longer made any sense. The first number indicates the male, the second the female. What this is all about, you will see in more detail in the course of this chapter.

After the "Pioneers" had not been able to hatch any chick for several years (up to and including 1999) and a brood failed again in spring 2000, I was convinced that they were too old. Maybe the "Pioneers" really were the Methusalahs of the Brook; perhaps they had occupied the central territory since 1981 and had fledged several young. If so, they would have been about 20 years old in 2000. Not impossible, but unlikely.

A little later in the season, at the end of April, I got the impression that a new pair was breeding in the northern area of the Brook. But, at the same time, the "Pioneers" had disappeared from their territory. Had they moved? That would have been something totally new.

The "Pioneers" were no longer calling; I was reduced to guessing like everyone else. Only some weeks later did I succeed in recording a unison call, which showed me the following: What I was hearing there were the "Pioneers", my pair No. 1, and they bred in the north. They guided two juveniles that fledged later. For the first time in probably 5 years!

I, the alleged and self-proclaimed expert on the cranes of the Duvenstedt Brook and their behaviour, had thought that they were too old and would no longer raise chicks or had separated. They proved me wrong.

In the following year, the "Pioneers" practiced a new variant: They bred in the original central territory, but after a few weeks, they guided their (only) chick to the northern part of the territory, just as they had done once before, when the grass grew faster than their young. Again, I only learned this from a unison call that the pair gave me "as a gift" (usually, when leading their young-sters, cranes almost never emit the unison call).

So, the "Pioneers" changed their behaviour even faster and adapted their breeding and rearing strategy much more variably to the requirements than even I was prepared to admit until then. Unfortunately, we have no informa-tion about why the broods failed for years. Apparently, however, the pair had made an effort to learn from the failures, not dully repeating the same proce-dure in the same place over and over again, but rather growing from these earlier experiences and making drastic changes.

One could now assume that, in 2002, the "Pioneers" continued in this way: breeding in the "Great Moor", and then leading the chick in the north of the territory; or breeding in the north, perhaps only when trying to breed again after the first attempt failed. This seemed to be the case, because, in the central original "Pioneer" territory, another pair once again appeared. The northern territory was (as always) initially unoccupied and only occasionally visited by cranes. But in March, something was different: The pair almost always stayed in the narrower core of the "Great Moor", hardly ever venturing outside, and the very first time I heard the first calls allowed me, without digital analysis, to come to a conclusion (because the male had an unusually high voice): It was not the "Pioneers", not pair No. 1. It became my pair No. 13.

But where were the "Pioneers"? Every now and then, I heard calls from individual cranes, and finally, I managed something that I had previously thought impossible: I heard, stored and analyzed calls that sounded like a pair's unison call, but were presented by a male alone, the female being either not present or silent.[9] This was something that I had experienced before with the "Sly Dogs" and that only I had heard, and that many experts at that time had thought that I had only imagined—here, I had now proven it to be real.

Finally, I managed to get a look at the calling crane and realized: He was alone! He was calling the male vocal part of the unison call as a solo. It was the male M1, the "Pioneer". I felt sad: So, a crane marriage had ended. The female had probably died, and the male was now looking for a new mate.

Suddenly, however, more and more life gradually emerged in this north-eastern part of the Brook, a small but beautiful patch of wet forest where nothing had ever really happened before. Another single male appeared, and I witnessed cockfighting: The two cranes paraded side by side, both handsome males, very large and strong. I called the second male M17. There was no

---

[9] Solo "duet" call: http://bit.ly/32YhXEg

female to be seen. However, just a few days later, I saw a single female stand-ing a bit off to the side.

It wasn't long before M17 was out of the race. He did not get the girl. M1 had conquered the new female, and thus she got the number F16. I hence-forth called the pair No. 1,16, and it settled down to live in the small wet forest in the north of the Brook, quite secluded, but well audible. I heard and digitally documented a few acts of copulation. I saw them together all through-out the season, actively calling, but they did not appear to be breeding. Perhaps the mate was still a little too young.

Sure, I suspected, pair No. 1,16 would try the following year to reclaim the beautiful old territory from the "evil" occupying pair No. 13. But it wasn't to be. In March 2003, I could, at first, not believe my ears, nor my analyses. In any case, No. 1,16 was not to be heard in their territory of the previous year. In the northeast, a pair was heard whose call did not match that of No. 1,16, unless there had been a huge vocal shift in the female. The male could be clearly identified: It was the male "Pioneer". Had he found another young female? For years, he had been together with the same partner, and then, sud-denly, a new female, and the next year a new one again?

For 3 weeks, I heard this call and could analyze it, but did not understand who the female was. Suddenly, at the end of March, the unison call was again that of No. 1,16 from the previous year! How could this be? The female could not have had a deeper voice with a different frequency pattern for 3 weeks and then "flip" back to last year's voice pattern. I began to doubt my analy-sis method.

So, I dug into my archives and looked at other, older analyses. I should have done that earlier, because it turned out that, in the previous 3 weeks, the old "Pioneer" M1 had been together with his long-time female F1! They had not been in their original territory, but rather in the territory that the Pioneer M1 had moved into with his new female, F16.

Wherever F16 had been during these 3 weeks, she was now back, together with her male M1, and F1, the old "Pioneer woman", was out again.

So, F1 was alive! This was actually my biggest surprise, because I had been convinced that she was dead and that her death had been the cause of M1's remarriage to F16.

Once again, I was proven wrong. The life of the cranes turned out to be quite different from what we had simplistically thought it would be. My inter-pretation was now this: There was no doubt that M1 and F1 were relatively old cranes, at least 11 or 12 years old in 2003. Since the previous year, how-ever, the two had not been together. In that time, M1 had, out of his own free will, chosen to be with F16; for reasons that were not evident to us, he was no longer together with F1, who was not dead, but, in any case, was apparently not to be seen or heard in the Brook in 2002. Who knows where and with whom she had spent the spring and summer of 2002?

In 2003, however, she was back. Having initially shacked up with her old husband, she was then either expelled by the younger (?) female F16 or left voluntarily, and the old Pioneer was back in the northeast territory with his new partner. They tried to breed again that year, but again without success.

Wasn't that crazy? It was hard to digest that cranes are not, by all means, unbreakably faithful to their partner for their whole lives, but apparently mate again even if the previous partner is still alive. In addition to my acoustic analyses, this is also supported by a few analogous observations from banding projects which ran several years later in Brandenburg. There are probably reasons for these separations that are somehow conditioned in the partnership. To put it simply and without anthropomorphizing: The harmony between the previous partners has been lost.

However, it was beyond even my imagination to observe a couple being back together for at least 3 weeks more than a year after a breakup.

After the separation, I noticed F1 two more times: Once, I saw and heard a single female in the south, whose call signature indicated F1. Then later, I heard guard calls from a male and a female: I concluded after the analysis that this was M17, along with F1. M17 had courted female F16 the previous year (2002) in competition with M1 and lost. However, I did not perceive a unison call between M17 and F1. M1, F16, the newlyweds, were not to be found in 2004; they did make an appearance in 2005, but not thereafter.

The new pair No. 13: In the early spring of 2002, a new pair took over the original territory of the "Pioneers", the "Great Moor". It received the number 13. Already, from a purely visual standpoint, I noticed that the male was not particularly large and the female was particularly small.

When I saw the power spectrum in front of me while analysing their strikingly high calls, I thought: I have seen something like this before, in calls from the Nienwohld Moor! This moor is about 10 km north of Duvenstedt Brook as the crow flies; I had recorded there for the first time in 2000, but didn't get around to doing so again in 2001. Actually, I had intended to check whether there was an occasional exchange of pairs or single cranes between these two areas, which would not have been a surprise given the short distance, although this had never shown up anywhere in any crane monitoring.

The Nienwohld Moor is a small nature reserve. It consists of a treeless central raised bog, some lower-lying peat-covered bog areas where there is now water and where numerous birch trees and various bushes grow, a small bog lake and surrounding green land, some wild and some extensively managed and partly grazed. Cranes mainly look for nesting sites in the high moor, although they sometimes move to the slightly lower rewetted moor areas.

So, I now looked up the power spectra of the crane calls from the Nienwohld Moor in my year 2000 archive and indeed found a female that also called at a very high pitch. So far, my memory was correct, but the shape of the acoustic spectrum was completely different. This was not the female that had now turned up in our area. And the male didn't match the previous profile either.

But I found another spectrum from the Nienwohld Moor, that of a male, clearly identical to the male of our new Brook pair No. 13, which had appeared here now, however, with a different female than at that previous time in the Moor. This was the first time that I could prove what was actually to be expected: Cranes or crane pairs living in the Brook know other places and change their territory over certain distances, and not only within the Brook. In fact, 10 km is no great distance for a crane, but such a change of breeding site had never been observed before, let alone combined with new mating.

Pair no. 13 was active in the Brook. Both were always in the area, calling dozens of times a day. Soon, I could identify them by ear alone, because they had such a characteristically high-pitched sound. They were copulating, but did not appear to be breeding. Male and female were always together, always in the narrower area of the territory, often in the extensive reed beds where they could not be seen but could be heard all the more intensely. Presumably, they were building nests, but apparently not laying eggs.

In 2003, it was the same show all over again: The two were always there, called often, but did not breed. Were they not ready yet? Were they too young? In 2000, M13 was at least 2 years old, so in 2003, he was about 5 years old or more. Maybe the female was 1 or 2 years younger. Is that why no brood succeeded? And how did this pair manage to defend this territory, which was the most coveted of all?

I waited for the year 2004—and what did I discover? M13 had chosen a new female (F23) and continued to occupy the "Great Moor". And F13 was with a new male (M24). This new pair settled west of #13,23. So, both previous mates were there, each had a new mate, and they were neighbours. How did this happen? Had F13, who already knew the Brook, told her new partner that there was a great territory to move into? That would again be—like the previously mentioned "Planner" female F2, which practically took over her old territory with her new partner M19—an example of the fact that it is not always the male crane—as is generally assumed due to ringing projects—who decides where a pair will look for a territory, but also sometimes the female, and this even in the immediate neighbourhood of her ex-partner!

If you as a reader now get the impression that, obviously, nothing is predictable with the cranes and every life story is different, then you have arrived in a very short time at a level of understanding that it took me several years to attain.

Only 1 week after this observation, Beate Blahy, Eberhard Henne's wife, called me. She wanted to make an appointment with me to make sound recordings of the crane calls in Brandenburg that year. I told her what I had just recently discovered.

"You won't believe it!" Beate shouted through the phone, laughing. "We banded a few adult pairs last year, and of course, stupidly, one of those males comes back from winter with a new female. We get annoyed—what's the point of all that effort if the female is then lost—but a week later, the female we ringed last year turns up—with a new male, and the two pairs seem to be taking possession of their territory not far apart!" This was a direct and independent confirmation of my observations. If I myself had still been a little unsure as to whether my story, which, after all, seemed rather farfetched, could be true, the last doubt had thus vanished: Females, too, can obviously determine which territory they want to occupy with their partner, and ex-partners may still live next door to each other.

Of course, I kept an eye on the two neighboring pairs. For 3–4 weeks, they lived side by side; I don't know whether or to what degree it was amicable. In any case, they were both occasionally heard.

But then, from the beginning of April 2004, pair No. 13,23 could no longer be heard. On the contrary, No. 24,13 occupied the whole of the "Great Moor", calling from all positions, from the eastern corner to the northwestern. The only other pair that they acknowledged was the old, very well-established pair No. 4 (the "Foxes") in the southeast. No. 13,23, however, had disappeared; this newly assembled pair with the original territory-owning male M13 had been driven away. Had M24 or the female F13 planned this long before?

But this was by no means the end of the lively story: Still at the beginning of June, i.e., rather late in the season, a new pair appeared, took over the territory of No. 13,23, the eastern, more open part of the "Great Moor", and proceeded to defend it. I arrived at the Brook on a Sunday morning and heard guard calls from a distance, which I immediately followed. Soon afterwards, I saw the pair. On later analysis, it turned out to be a new pair; I called it No. 31. I walked on, losing sight of the pair, but all at once, I saw it take off. It then landed in the meadow right next to where I had seen it before. Only a little later, I heard soft copulatory sounds and began to record them. The sounds were coming from the exact direction of where the pair had landed.

It doesn't always happen, but quite often, pairs will call in duet directly after copulation—and so did these two this time! How nice, because now, I could compare the signature of the warning and unison calls right away.

Over the course of the next few weeks, I still heard the pair occasionally copulating, and eventually, they seemed to be attempting a brood, but this was apparently abandoned after a short time. If a nest was built, and I have not misinterpreted my observations, this appeared to be in the territory of No. 4 (the "Foxes"), which had retreated farther to the east and south.

So, there was another new pair that had found a home in the Brook. In 2005, it could not be found; in 2006, it was breeding in the same area as in 2004. In 2007 (when I was abroad a lot due to work and rarely had opportunities to go to the Brook), I was unable to find it again.

Pair No. 12: In addition to the most beautiful central area of the Brook, the north also saw a new pair appear in spring 2002. At first, it stayed far outside; I heard (and digitally preserved) calls from the direction of Jersbek. Soon, however, it moved into the area in the north of the brook, where the "Pioneers" had incubated or led their chick in the 2 previous years.

If I had not been able to record and analyze the calls of the pair with the young offspring at that time, despite the late season, I would not have been able to recognize that it was not, in fact, the "Pioneers" in 2002, but a new pair. By my counting, it became no. 12.

No. 12 was a very active pair. It explored the entire area and occupied a large territory. This territory had been avoided by the cranes—with the exception of the "Pioneers"—for two decades. Two meadows in this territory had seen less intensive use by the farmers in the recent past and were now slowly going wild. Fortunately, their drainage was functioning less and less well, and the wild boars were digging up large areas. This made these pastures quieter and more attractive to the cranes, and pair no. 12 had recognised and taken advantage of this fact.

In the second half of April, I assumed that they were breeding. You hardly ever saw them anymore. Gradually, however, I realized that they were not breeding, they had just become very shy. They no longer lingered in the large pastures, only in the background, making them hard to spot through the sparse birch woods. Eventually, I came to the conclusion that the pair was moulting. This would confirm that the birds were at least 3–4 years old. Accordingly, we could expect a brood for the following year, 2003.

And so it came to pass. The two kept their breeding site well secret. It was only after a while that I happened to notice that it was in a place where I had heard them calling in unison the year before. Presumably, they had already inspected it as a potential breeding site then: Amazingly, the nest was only 50 m from the trail and 150 m from the road surrounding the north of the Brook. How could we have missed the nesting and the subsequent replacement of the partners for breeding?

After the usual period of 30 days, pair No. 12 appeared in one of the pastures, apparently guiding a chick. About 2 weeks later, the chick had been lost, an event that is nothing unusual for young, inexperienced breeding pairs. It was not until 2006 that the pair managed to fledge their offspring.

It should have become clear by now that cranes also lead a rather eventful life and are not necessarily spared from the strokes of fate. In this context, the **male M6** comes to my mind. I got to know him in 1997 as part of a youth group—even then, I was struck by the characteristically rough voice of this animal. In 1998, this male started breeding with his female F6 (I called the pair the "Rattlers"). They stayed together for 2 years, without successfully breeding (maybe one of them or both were still too young). In 2000 and 2001, M6 managed to breed with another female, F11, in a different territory. In 2002, he was again successful, this time with female F14. He also bred successfully there in the years 2004 to 2006 with female F28. Whether these two were also successful in 2007, my last year of recording for the time being, I could not determine, but at least they were together (M6 had probably become calmer and more consistent over the years without losing his flexibility, because he occupied five different territories in succession, more than any other crane in the Brook, ultimately spending several years in the Rade Forest directly next to the Brook).

The pair known as the **"Foxes" (No. 4, M4F4)** was one of the few pairs that was present in the Brook for nine of my 10 years of observation. Male M4 was the only male crane to appear in each of the 10 years of my bioacoustic monitoring. In 2005, however, it was not with its previous female, F4, but with a new partner, F32. However, this relationship lasted only one year (and was also unsuccessful). In 2006 and 2007, M4 reappeared with F4, and they even had another fledgling in 2006. Isn't that interesting and a terrific confirmation of the other equally surprising observations? None of these are simply rare exceptions! (By the way: The answer to your question why I called this pair the "Foxes" can be found in the chapter "Can cranes think strategically? More amazing observations".)

Beate Blahy and Eberhard Henne reported on their observations of banded crane pairs in Brandenburg at the European Crane Conference in Arjunzanx/France in December 2018. They succeeded several times in banding adult mated cranes when they were moulting. Their conclusions are very similar to mine from bioacoustic analysis: They too say that there is no fixed pattern in the life of cranes. You find firmly and stably mated cranes; you find those that more frequently seek another mate; and the territory owner is predominantly, but not always, the male. Re-mating is the rule rather than the exception, as it is in my Brook crane society.

A particularly interesting example from Brandenburg is the crane female "Rosa": Rosa was banded in 1997 (she was estimated to be 6 years old at that time). Up until 2003, she only changed her partner once. In 2003, Rosa was re-banded; on this occasion, her new partner was also ringed and given the name "Bernhard" (this was because I had helped with the ringing and had recorded the pair's unison call several times). The pair stayed together until 2013, when Bernhard disappeared. Rosa then mated with "Jochen", whom she was taller than. By 2017, Jochen had also disappeared. Rosa was then seen together with two different males, but each time only for a few weeks. In 2018, Rosa returned from winter vacation alone; she was with another male soon after, but they did not breed. In the 24 years of her life so far until 2018, Rosa has had seven different mates and, with them, eleven fledged young birds; she is the territory owner. For cranes, it seems to me that mate changes are much more regular than we had previously believed, and that "mating for life" is the exception (Graph 7).

This does not at all seem to be exclusive to cranes. Lloyd Spencer Davis, a New Zealand penguin researcher in Antarctica, determined that these flightless birds have very dynamic relationships and sex lives. He learned of homosexual behaviour, prostitution, rape and mate changes, publishing his discoveries, only to find out many years later, that the world's first penguin researcher, George Murray Levick, had discovered and documented the same things 83 years before him. But Levick had not dared (and was not allowed) to publish them; he had even written his notes in secret codes. The common morality in the Victorian age in Britain, as well as his own moral code, prohibited uncovering the secret sex life of penguins. Davis summarized his own findings from his research on Emperor Penguins as follows: "Emperor penguins are not the epitome of love, but rather the patron saints of divorce."[10]

The divorce rate of penguins—although different depending on species, with the highest rate being found in Emperor Penguins—seems to be on the same order of magnitude as that of cranes. Also, homosexual behaviour and "speed-dating" is obviously not impossible for them (cf. the last section in the attachment). Only, in contrast to penguins, which can be openly observed in their breeding grounds (provided you can get a research position in the Antarctic and do not mind spending endless hours in the cold while nothing worth noting happens until very rarely something is happening), cranes mostly stay in hiding. Nevertheless, I have uncovered some of their secrets.

When describing new surprising observations with this species to friends and acquaintances who could not believe such stories, I sometimes jokingly said: "Why not? Cranes are also only human!"

---

[10] Translated by myself from the German edition: L. S. Davis, "Das geheime Liebesleben der Pinguine", dva 2021; original english edition "A Polar Affair: Antarctica's Forgotten Hero and the Secret Love Lives of Penguins", Pegasus Books 2019.

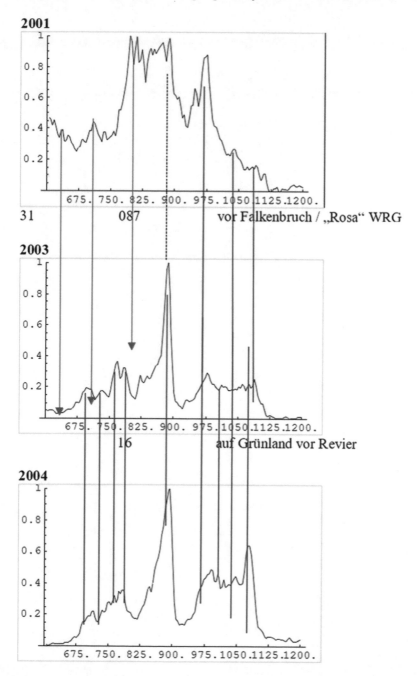

Graph 7 The banded female crane "Rosa" in 2001 and 2003 paired up with the same male, in 2004 with a new male

Off into the Wide World: Asian and American Crane Species Are Calling Me

As early as 1998, word began to spread through the world of crane experts that I had developed a method for individually identifying cranes by their voices.

When I reconnected with George Archibald, then president of the ICF, at a conference and told him about the results of the evaluation of Eberhard Henne's recordings, he immediately wanted to try out the new method on other crane species. So, he invited me to Baraboo, USA, where the ICF is based.[1] In Baraboo, I was to record unison calls from the most critically endangered crane species, which are bred there for reintroduction or for the preservation of genetic material, and to investigate whether individual "banding" via voice analysis is also possible with these crane species. Following a business trip to the USA, I spent several days in January 1999 in "Crane City", the breeding center of the ICF.

George Archibald is one of the two founders of the International Crane Foundation. It has become the world leader in crane conservation and crane research in its now 50 years of existence, and it has done so with purely private funding—thousands of members contributing, and more importantly, major sponsors donating larger sums. Although a native Canadian, George ran the organization in true American fashion, with gala dinners and campaigns handing out colorful crane buttons and balloons. This would elicit a frown from many a conservationist of German descent. In the meantime, George, now a very active retiree, is a "Senior Conservationist" and "Board Officer". His successors have managed the Foundation as successfully as he did. I

---

[1] cf. https://www.savingcranes.org/

© The Author(s), under exclusive license to Springer Nature Switzerland AG 2022
B. Wessling, *The Call of the Cranes*, https://doi.org/10.1007/978-3-030-98283-6_9

consider myself fortunate and honoured to have been able to regard George Archibald as my friend for a long time.

There are 15 crane species worldwide, of which practically the only one that makes its home in Europe today is the Grey Crane (Eurasian or Common Crane, *Grus grus*). In southeastern Europe, there are still small populations of the Demoiselle Crane (*Grus virgo*), which used to be found over almost the entirety of Europe. Today, Demoiselle Cranes live mainly in the southern former Soviet Union, Mongolia and northern China.

More than half of the world's 15 crane species are endangered, some of them very severely. The most endangered species is the American Whooping Crane (*Grus americana*). Similarly problematic is the situation with the Red-Crowned Crane (*Grus japonensis*), which now again numbers 1900 specimens in Japan and about as many on the Asian mainland in China and Russia, consisting of two populations, one west of the Khingan Mountains (China, Inner Mongolia Province), about 500 birds in number and declining, and another east of these mountains in the Amur region, about 1600 birds and increasing.

The so-called "eastern" population of the Siberian Crane (*Grus leucogeranus*) has about 4500–5000 birds, a number that, fortunately due to dramatic conservation efforts in China and Russia, shows growth tendencies. The former "western" population is similarly practically extinct—only one crane is still alive. The reason is most probably illegal shooting on the migration corridors to their wintering range in Iran and India.

Not much better off are the Black-Necked Cranes, White-Naped Cranes, Hooded Cranes, Whattled Cranes and Blue Cranes. I first saw wild Black-Necked Cranes (*Grus nigricollis*) in June 2016 in China, at QingHai Lake, at 3200 m above sea level. While my partner and I laboriously climbed the few hills from which we could look with binoculars into the vast plains in search for the cranes, I made it clear to myself: These cranes (like bar-headed geese) apparently fly without complaint at up to 10,000 m above the Himalayas to the south to reach their wintering grounds. They don't almost break down like we do when they want to climb 50 meters (Fig. 1).

Cranes are endangered wherever humans drain wetlands, and thus cause the birds' nesting and feeding areas, their meeting and resting places, to disappear. A "side effect" of this overexploitation of nature is that, in the long term, we are draining ourselves of water, thus preventing the regeneration of groundwater. In addition to the cranes, many species of plants, insects, amphibians and birds are threatened or already extinct because of the massive destruction of wetlands.

Fig. 1 Two Black-necked Cranes and two Eurasian Cranes are exchanging unison calls across the species boundary. (© International Crane Foundation, Mike Endres)

Unfortunately, knowledge of the importance of wetlands for the climate, for example, as water reservoirs, or of peatlands as active CO_2 reservoirs (a relatively recent finding), has so far had little influence on human actions in practice.

Protecting cranes means, first and foremost, preserving or restoring their habitats, i.e., the entire biotope systems of wetlands and their surroundings, together with their inhabitants. Thus, crane protection initially means the protection of marshes, wet forests and wet meadows—diverse, lively and fertile landscapes that are habitats for many other species, whose continued existence is ultimately inseparable from the survival and further development of humankind.

While the habitat loss described above is the most damaging factor for cranes worldwide and the main cause of their endangerment, there are two other causes, hunting and poaching, which should not be underestimated. Our local Grey Cranes have been wiped out in the UK due to hunting and have only managed to regain a foothold there with a few pairs in recent years. Sandhill Cranes, for example, were intensively hunted in the USA until the early twentieth century, so that, by 1920, there were only 26 pairs left in Wisconsin.

Now, hunting having been restricted and partially banned, there are currently more than 12,000 Sandhill Cranes again living in this state, which is rich in lakes and wetlands. In total, the number of individuals in Canada and the northern USA has been restored to about 700,000.

Whooping Cranes were hunted, in part for food, in part for their magnificent white feathers, which were a popular adornment on the hats of the wealthy American women of the East Coast cities, until there were almost none left. In addition, the draining of their breeding habitats (wetlands) was a crucial reason for their decline to the point of becoming almost extinct. In 1946, the low point was reached, with just 15 individuals remaining.

Today, cranes are only hunted in the Middle East and some countries in Asia, and this is one of the greatest threats to the last Siberian Cranes. Regardless of this, one of the goals of the ICF is to ban the hunting of cranes everywhere and to have this ban efficiently controlled. Nevertheless, in spite of the ICF, whose American center is in Wisconsin, having taken a stand against the hunting of cranes, each year, many thousands of Sandhill Cranes are legally hunted in many states of the USA, as well as in a couple of Canadian provinces. The Republican Party of Wisconsin has recently (October 2021) proposed a bill to again allow the hunting of Sandhill Cranes in this state.

However, efforts against hunting is only one of the ICF's work areas. Worldwide, the foundation has about 120 salaried employees, supports long-term conservation projects in 12 countries across Asia and Africa, and maintains a network of crane specialists in 60 countries around the world. About half of ICF employees are based in Wisconsin, in addition to countless volunteers, dozens of interns and individual foreign guest researchers on a site about 110 ha in size, a small paradise. Nearly 20 staff work in East Asia, ten in Southeast Asia, and another 20-plus in Africa. It is here, and in the research and conservation projects underway around the world, that the money is needed, with an annual budget of over eight million dollars.[2]

The more than 25,000 visitors each year (who contribute to the foundation's income with their entrance fees) are offered an attractive programme. The ICF has thought of everything, including electric trolleys for people with limited mobility. In the Foundation's public "zoo", you can see and learn about all 15 of the world's species of cranes, the only exhibition of its kind on the planet. All of the crane enclosures feature wetlands, native vegetation, and interpretive signage, including colorful murals that depict the natural places where the cranes occur in the wild. Wattled cranes (Bugeranus carunculatus) are housed in a vast open enclosure of native prairie and wetland, and two Whooping Cranes have a large pond where they forage up to their bellies in water. The cranes seem to like it there, as several pairs that had never laid eggs

---

[2] cf. https://www.savingcranes.org/about-icf/financial-information/

Fig. 2 A Whooping Crane forages in a pond at the International Crane Foundation's crane zoo which is open to the public. (© International Crane Foundation, "Whooping Crane Wading in Exhibit Pond," Crane Media Collective, accessed November 15, 2021, http://gallery.savingcranes.org:8082/items/show/22436)

before living there began building nests and breeding in the "open-air theatre" as early as their second season (Fig. 2).

On almost 4 km of trails, through 40 ha of forest, wetlands and prairie, visitors can get to know the original vegetation of this region, which has been actively restored on former farmland over the past 50 years. The aim is to show visitors that the preservation of habitats and biodiversity in self-regulating biotopes is important and that it is not just about the conservation of individual species. Anyone who walks through the prairie grass, some of which is taller than most men, and along the narrow mowed paths, who breathes in the smells and listens to the whispering of the grasses and the chirping of the crickets, or who watches the countless birds, will understand this message. On my first visit, I could hardly believe that a purely private initiative had created all of it.

Not open to the public is "Crane City", the approximately 30-ha facility with the breeding enclosures. There, George and his colleagues began breeding endangered crane species soon after the foundation was established. For the first time in the world, they succeeded in breeding Siberian Cranes, another extremely endangered species. There is a fascinating story behind George's success in transporting the eggs of Siberian Cranes from the (at that time) Soviet Union to the US; he personally attended to all aspects of the move himself, and somehow managed to get everything officially approved by both sides. What an achievement!

Whooping Cranes were originally allowed to be bred by only one state institution in the United States, Patuxent. The ICF became the first private institution to be licensed to breed alongside the Patuxent Wildlife Research Center. The Patuxent Center had to give up its crane breeding in 2017, due to the austerity measures of the Trump administration, and thus the cranes there were mostly handed over to the ICF.

Breeding alone would not help the Whooping Cranes. The real step towards saving them would be to release those Whooping Cranes raised in captivity into the wild.

Beginning in 1975, initial attempts were made to reintroduce Whooping Cranes by putting Whooping Crane eggs from Patuxent and ICF into the nests of (non-threatened) wild Sandhill Cranes. These attempts failed and were abandoned. The eggs were successfully incubated, the ignorant Sandhill Cranes raised the chicks and also taught them how to fly and migrate—but the white Whooping Cranes refused to join their equally white conspecifics: They apparently did not find them sexually attractive, and were only interested in the grey-brown Sandhill Cranes upon which they were imprinted. However, they could not form a fertile partnership with the latter; the species barrier prevented it.

In 1993, a program was begun to establish a reintroduced population of Whooping Cranes in Florida. The "Whooping Crane Recovery Team" (WCRT) had been preparing this action for a long time. First of all, the method of rearing had to be changed, because it was important that the cranes not become imprinted on humans, but rather on conspecifics. In 1995, the team began to develop an "isolation rearing method" in which the young cranes never saw humans directly. Instead, humans appeared—if at all—camouflaged by white "crane plumage" capes and fed the young animals with special spoons modeled after a crane's beak.

Nothing was left to chance. Even the transport was specially organized: The selected young animals from Baraboo or Patuxent were not sent on a commercial airplane, but on a private chartered airplane that was specialized for such operations.

In Florida, the animals were kept in quarantine for weeks and regularly examined. Gradually, they were accustomed to freedom and to the already free-living cranes. They were to join their groups and spend the night with them in the wetlands.

Nevertheless, the loss rate was quite high, at around 50%. Alligators and bobcats found easy victims in the inexperienced animals. By 1999, 75 Whooping Cranes lived in Florida, with about 30 young cranes being released annually.

Gradually, the first pairs formed and began to defend territories. But no breeding was ever observed at first—until finally, in 1999, the wonderful news reached experts all over the world by e-mail: A first one, then a second and a third pair had laid eggs and were breeding! But shortly afterwards, the euphoria faded, as one clutch after another was lost. Two failed due to drought: The eggs were found broken; it remained unknown which animals had done this, and it could not be ruled out that it was the cranes themselves (in captivity, this happens quite often when inexperienced adult Whooping Cranes try to hatch eggs).

A third clutch was flooded during a torrential rain; it was assumed that the cranes left the nest when they felt the cold from below. Shortly thereafter, the clutch was found. One egg was still intact; an attempt was made to hatch it artificially, but without success. It was too late.

Already at the end of the 1990s, when I was involved in the Whooping Crane program, people were questioning the sense of this reintroduction project. It was subsequently halted in 2007, because the loss rate was simply too high. Even before that, focus had shifted to a project in which a population in Wisconsin would be established through reintroduction, one that would hopefully migrate to Florida in the winter. I was to be involved in this program, albeit marginally, namely, in the identification of individuals from the last wild population, and for laying the groundwork for their bioacoustic monitoring. For this, I was to provide unison calls of Whooping Cranes, which I wanted to record in Baraboo; later during monitoring, they should sound via megaphone to provoke duet responses from wild cranes.

I started my new task on the very first day after my arrival in January 1999. At minus 25 to minus 30 °C, with stiff frozen fingers, I spent two days making recordings of unison calls of various crane pairs in "Crane City", especially of Siberian, Whooping and Red-Crowned Cranes. It was on these species that I wanted to focus my future work. In the following months, I also visited the German Walsrode Bird Park (which kept and bred some crane species), the state breeding facility in Patuxent near Washington, and the Calgary Zoo in Canada. There, I recorded more crane pairs. The analyses on the computer showed that the calls of these crane species are also individually different and characteristic.

George Archibald immediately thought of half a dozen areas and projects around the world that he wanted me to visit to record crane calls and build new databases using my bioacoustic method.

This was a challenge that I would be able to balance well with my business travels. Since I was no longer heading the crane protection programme in the Brook, time reserves were freed up for my international bioacoustic research as well. At home, until 2007, I made recordings of crane calls in the Brook and continued my behavioral observations there. In addition, I made and analysed numerous recordings over several years in Mecklenburg, Brandenburg and the Nienwohld Moor.

When Bill Lishman, George Archibald and I met, it represented three very different characters coming together, each one "crazy" in his own way: an artist, a crane foundation director, a chemical entrepreneur. The three of us each funded our work through purely private sources, independent of government money (although still dependent on government permits), and we bridged the differences in our views about the best way to protect the cranes through open, intense, and humorous discussions, until the three of us were convinced that we had found the best solution.

Bill Lishman[3] is an example of the power that can come from the initiative of an individual. He is best characterized by this quote from a program on Canadian television: "It would be a loss to the world if Bill Lishman took permanent employment anywhere." Bill was unconventional in every way; compared to him, I am a dead-beat philistine. He achieved a level of crazy that I would have never thought possible in any of my acquaintances. So crazy, in fact, that he didn't even live in a "normal" house, but in a round cave house of his own making, with not a single corner in it, and no windows—for the house was entirely underground, except for the entrance and the ceiling-light windows.[4]

Bill, who passed at the end of 2017, is one of the pioneers of ultralight (UL) aircraft technology. As a youth, his dream was to fly with wild geese. He fulfilled that dream as an adult in 1988: After many failures with various types of home-built aircrafts, he finally learned how to train his geese to follow him in his plane into the air. He financed this hobby project with proceeds from his artistic work. It was he who (inspired by George Archibald) suggested to crane conservation organisations in Canada and the USA in the early 1990s that they should guide young cranes intended to be released into the wild from a northern breeding area to a southern wintering area, thus allowing them to learn migratory behaviour; and to use UL airplanes for such a migration guide project. Shortly thereafter, Bill and his friend Joe Duff, a photographer, established a foundation for this purpose with the catchy name "Operation Migration." It became the central operational partner of the "Whooping Crane Recovery Team" (WCRT), which consisted mainly (apart from the ICF) of government institutions. Bill, a steel artist, and Joe, a

---

[3] Bill died on December 30, 2017 at the age of 78 years.
[4] https://bit.ly/2RuugSB

photographer, were only qualified for the key role in this project because they had already flown with wild geese. They funded the launch of "Operation Migration" with the proceeds from the film *Fly Away Home*,[5] for which they and their troupe trained the geese and performed the flights. Bill, the engineer of this innovative ultralight aircraft, piloted the craft as a double in the film, and can also be seen as a pilot himself and as a motorcyclist who picks up Amy.

George Archibald, the restless founder and driving force, fundraiser and scientific inspirer of the International Crane Foundation (ICF), is completely different, and yet in some ways, the two were comparable. He, too, is an example of the power that an individual's restless initiative can generate. He founded the Crane Foundation in 1973 while still a student, along with his friend Ron Sauey (who sadly passed away all too soon). The two of them—and soon George alone—worked tirelessly to get more and more people involved in crane conservation and nature, biodiversity conservation and to raise funds. George has become the world's leading and most respected international crane conservationist and researcher over the now many decades.

His 2016 book "My Life with Cranes"[6] contains an impressive number of adventurous stories from his life. The most famous, and at the same time most unexpected, story is the one in which George dances with a female Whooping Crane. It became known throughout the US because George was invited to Johnny Carson's *Tonight Show* in 1982 and described his successful dance with a crane there.[7] The female Whooping Crane, "Tex," hatched at the San Antonio Zoo in 1966, at a time when there were only 44 Whooping Cranes in the wild and 15 at Patuxent. Tex was the first chick hatched by a pair at the zoo. Tex was brought to Patuxent after 3 weeks of care at the zoo director's home and raised there. Thus, this crane had been irrevocably imprinted by humans, and for years, it was not possible to mate Tex in Patuxent with the formerly wild crane "Canus" or any other male. What was the problem?

The Whooping Crane is the most endangered crane species. In the winter of 1945/46, there were only 15 specimens left. With an extraordinary effort, a few conservationists succeeded in protecting the wintering grounds of the Whooping Crane near Aransas, Texas. It wasn't until 1954 that the last

---

[5] https://en.wikipedia.org/wiki/Fly_Away_Home

[6] George Archibald: My Life with Cranes—A Collection of Stories. Baraboo: International Crane Foundation 2016.

[7] To my knowledge, there is no complete online video copy of this part of the show, but there is a video compiled by the ICF that includes clips from the show: https://bit.ly/2S0cgUU (last visited Sept. 23, 2021); at https://vimeo.com/58841603, as well as a video with original footage and one about George on the occasion of an award he received in 2006. The following video also contains excerpts from original recordings of his dances with Tex and from the *Tonight Show*: https://www.youtube.com/watch?v=qyzJCvZoI44 (last visited Sept. 23, 2021).

Fig. 3 A just 4 weeks old parent-reared Whooping Crane chick in the International Crane Foundation's crane breeding center. (© International Crane Foundation, "27 Day Old Parent Reared Whooping Crane Chick," Crane Media Collective, accessed November 15, 2021, http://gallery.savingcranes.org:8082/items/show/22809.)

breeding ground of Whooping Cranes was discovered in Canada's far north, and it was protected with the creation of the Wood-Buffalo National Park. When it was realized that in each case only one young survives from a clutch of two eggs, people began to remove individual eggs and breed the Whooping Cranes in captivity (Fig. 3).

In 2020, thanks to massive protection efforts, the last wild population had grown to more than 500 individuals.[8] This is by far not sufficient to save this species, especially considering that it has a history of going through a genetic bottleneck, and therefore the Whooping Crane remains extremely endangered. An epidemic in the breeding area or a single oil spill in the Gulf of Mexico could wipe out nearly the entire species. This is another reason why captive breeding is maintained. It is hoped that this will also slowly regenerate as rich a gene pool as possible for this crane species (which is extremely

[8] Anne Lacy, ICF, Presentation at the European Crane Conference 2018 in Arjuzanx/France (Proceedings 9ᵉ Conférence Européenne Grue Cendrée 03–07 Dec 2018, Réserve d'Arjuzanx (eds), and George Archibald, personal communication.

homogeneous due to the low population of only 15 cranes): Genetically unique birds are kept alive and made to reproduce. One such specimen was Tex.

George Archibald made it a personal matter of the heart to get Tex to lay an egg. His effort lasted 7 years, during which George was with Tex more or less intensively each spring, dancing with her (which, if you know George, is hard to imagine). Tex was "in love" with George; she would not accept other people. George called his dances with her "Texercise", because it was strenuous exercise! And eventually, Tex laid her first egg, and later more, but all of them either failed to fertilize or had too soft a shell and broke.

In 1982, George wanted to make one last, especially intense attempt. He built a cabin near Tex's large enclosure, moved his office there, and stayed overnight from March to May. The operation was coordinated with Patuxent. The head of crane breeding there, George Gee, regularly sent crane sperm, which George then introduced to Tex after one of the dances.

Eventually, Tex laid a fertilized egg, which, after many more complications, produced a chick—the male Whooping Crane "Gee Whiz", named after George Gee. It was this success that got George Archibald on the *Tonight Show*, where 22 million Americans watched him—it was an incredibly effective breakthrough in public awareness of the Whooping Crane conservation program and nature conservation in general.

George has been instrumental in improving the breeding of Whooping Cranes based on his experience with Tex. He is unstoppable in his commitment to the cranes. He succeeded in proving that the Red-Crowned Cranes on the Japanese island of Hokkaido also had a breeding ground there, and he even found it. After this, he was able to convince the Japanese government to do more to protect these cranes on Hokkaido. I don't know how he was able to get into the hermetically sealed demilitarized zone between South and North Korea in 1974—but he got in and discovered the Japanese Ibis there, which was thought to be extinct in Korea, as well as there being only 20 of them left in Japan at the time. This success made him further well known in Asia. He also managed to extract a promise from then-Indian Prime Minister Indira Ghandi at a personal meeting to more effectively protect an endangered crane sanctuary in India, ban hunting and control the ban.

In Baraboo, George started a breeding program for the extremely endangered Siberian Crane by persuading the Philadelphia Zoo, on the one hand, and the owners of the Walsrode Bird Park, on the other, to transfer their Siberian Cranes, each of which lived alone, to Baraboo. The male from Walsrode was 71 years old when he arrived in Baraboo; although he and the Philadelphia crane mated, the eggs remained unfertilized. So, George asked the Hirakawa Zoo in Japan to send their Siberian Crane, which was also solitary, to Baraboo. However, this supposed male mated with the now 72-year-old Siberian Crane-Methusalah, an effort that also failed, although the new bird, now recognized as a female, did lay eggs. Only with another Siberian Crane from the Walsrode Bird Park in Germany did the first breeding succeed.

George and his friend Ron developed personal relationships, even in the then-Soviet Union, in the interest of the cranes. George was respected in the USSR as he had succeeded in the world's first captive breeding of a Siberian Crane. Because the first warming tendency in the Cold War led to an environmental agreement with the US, cooperation was elevated to an official level. After complicated negotiations, George personally received permission to transfer wild Siberian Crane eggs from the Soviet Union to Baraboo, which greatly improved further breeding there.

George was also one of the first to seek intensive contacts with naturalists and conservationists in China. In the early 1980s, ICF received a pair of Black-Necked Cranes from China in exchange for seven pairs of cranes of other species, which were given to a newly formed sanctuary (the ZhaLong National Park) in Northeast China located at an information center. In 1985, George succeeded in discovering the first Siberian Cranes at a lake only shortly before recognized as a wintering area for cranes in China (the PoYang Lake) during an extremely challenging expedition: This made the migratory movement of these cranes much clearer. George and his companions had also found 1000 more Siberian Cranes than had previously been estimated as the population of this species.

In the Australian bush, George searched for and found Sarus Crane nests. He researched cranes in Africa, Vietnam, and Thailand, in Mongolia, in Botswana (where he got stuck in a jeep in the open at night while looking for cranes and was forced to spend the entire night alone, using a fire to protect himself from being trampled by elephants or eaten by lions), and even in North Korea, where his work was to have a very different impact many years later (chapter "Cranes Are Subjects. A Plea for More Modesty and Respect for Nature"). Nothing is too burdensome for him, nothing too dangerous. At the same time, he never tires of fundraising, PR, negotiating with governments and the UN. I consider it a great honour that, for one of his booklets, in which he presents each crane species on the basis of a small incident, he has chosen for the Eurasian Grey Crane an incident that I myself observed.[9]

[9] George Archibald: Memories and anecdotes about the 15 Species of Cranes. Baraboo: International Crane Foundation 2018, p. 16/17. In this, George reproduced my observation of the pair the "Foxes", which I will describe later in this book.

Encouraged by the first good results of bioacoustical observations on Grey Cranes, George and I wanted to propose to WCRT that we make unison call recordings of wild Whooping Cranes, and thus begin broad monitoring. These calls and vocalizations, their analysis, and the best unison calls on CD were to be the basis for recording the calls of wild birds on their wintering grounds in Aransas. Because such an effort had to be approved by the entire WCRT leadership team and then by state agencies, and also required a tremendous amount of paperwork, the project had to be recommended by a full WCRT meeting of sorts.

I was nervous. Would the American team of experts entrust one of their projects to an amateur crane researcher from Germany (who was actually a chemical researcher and entrepreneur) without any experience with Whooping Cranes? To introduce myself and my work so far, I had prepared a presentation of my previous bioacoustic work.

Before my turn, the team discussed many other scientific and technical topics. I began to feel more comfortable in this group, and I was impressed. These people were professional, focused, comradely, and dedicated. Finally, I reported the results of my bioacoustic analysis to date, my initial analyses of calls in the Brook and in the Schorfheide in Brandenburg, Germany, calls of Whooping Cranes from the Baraboo and Patuxent breeding stations, and those of Red-Crowned Cranes on Hokkaido, which showed that individual recognition, and thus monitoring, of wild Whooping Cranes was possible. The lecture was a complete success. Now, I was accepted, almost like a long-time member of the team. And to my delight, moreover, my reflections on the behavior and intelligence of the Common Cranes that I had observed were received constructively and with interest.

We discussed the high loss rate of Whooping Cranes being released into the wild in Florida, many of which all too quickly became a victim of predators, mostly bobcats. I wondered why no one had thought about the fact that the juveniles had no experience at all with other animals before being released into the wild; after all, they had no parents to show them what dangerous creatures to be wary of. I asked if there was not a need to develop a "recognize and avoid predators" training program for young cranes. The idea met with general approval, and the very next morning, someone was distributing copies of articles on the subject.

A significant part of the discussion revolved around the reintroduction of Whooping Cranes to Wisconsin, and in particular, how to show the cranes that were to be released there the way to their wintering grounds. Bill Lishman explained the concept of "Operation Migration." His preliminary study with Sandhill Cranes had shown that it was possible to get cranes to fly behind ultralight airplanes. But after arriving in the target area, the young cranes did not want to stay among their own kind in the wild, but either followed the trainers and pilots or flew to market places in villages, to kindergartens, and even once to a prison yard where prisoners were exercising. If this were to happen in a similar way with Whooping Cranes later on, it simply would not be possible to release them into the wild.

I timidly raised my hand and asked, when given the floor, "What kind of sounds do you make while raising the chicks, while training the young cranes?" They claimed that, other than playing the sound of an airplane engine and a few environmental noises for the young cranes to get used to, they didn't use any sounds. I said, "I find that hard to believe; after all, you'll want to lure them to feeding areas and elsewhere!"—"Yes," they then admitted, "we make 'prrr' and 'drrr' sounds, like crane parents do." I objected, "But those aren't crane sounds, they're human sounds. Cranes, after all, have no vocal cords, no lips and a completely different sort of tongue to make the 'prrr' sound. That's why your lure sounds like a human voice, and that's how you imprint the chicks and the juveniles on humans."

Concerned, thoughtful silence followed. It was immediately clear to them that my conclusion was correct. "But what can we do?", they asked. My suggested alternative was to record the appropriate sounds of real, preferably wild, Whooping Cranes, digitally save them, and then use only those calls. The experts pondered this idea. How could it be done practically? I explained, "You use hardware like I do and sneak up on the cranes, preferably on their wintering grounds, because the breeding grounds are too swampy." That way, one could record those soft sounds at close range. This was, in my view, a logical extension of the project George and I had proposed to WCRT for the sonagraphic monitoring of Whooping Cranes; my plan was to train them so they could make the recordings themselves, after which I would conduct the analyses in Germany. The frown lines of the conferees, which were getting quite a workout, communicated to me the question, not asked aloud, "How on earth is that going to happen?" Finally, George asked, "Bernhard, can't you do it for us?" I hesitated. "But you want to start the monitoring with bioacoustics, I'll show you how to do it, and at the same time, you can do the other recordings, can't you?" George: "I think it's better if you do it all. You know how to record cranes, how to save and process the calls and so on—help us out, do it yourself please, will you?" Now, I was sitting there, having asked only one question, and the result was that I was supposed to procure the answers myself? My brain cells were going into overdrive: "Can I even manage this? What do I need to prepare for this? Can I fit it into my business, the chemical research and business I do for a living? I don't even know the area.

What about my family time? I don't know Whooping Cranes at all. I can't do this. Say, 'I'm sorry, I can't.' But I've never had such an extraordinary and important project proposed to me. Say yes. Don't pass up this chance for such a wilderness adventure!" And ultimately, my mouth was quicker than the conscious part of my brain: "Yeah, okay, I'll do it." Thus began my involvement in this great project to save Whooping Cranes.

My expedition was to take place in December 1999. To be ready, I had to prepare the various call recordings so that I could play them over a megaphone in hope of getting the Whooping Cranes to call in the wintering grounds. In addition, I had to record other (very quiet) vocalizations of Whooping Cranes there in the field to be used in rearing and flight training. My planned expedition to Texas had further goals, because, before a round of flight training with Whooping Cranes in autumn 2001, a final rehearsal with Sandhill Cranes was to be conducted first in autumn 2000. So, with these tasks, it turned out that I was not at the end of my crane research, but actually just at its beginning. At this point, I knew exactly nothing about crane communication or "language". I didn't even know if they had such a thing. I had only the impression that some calls have certain meanings or are emitted in certain situations, such as the unison call when defending a territory. But cranes also call in duets in other situations, for example, after copulation, when no other crane pair is to be seen far and wide. Do the calls have a different meaning then? Or do they actually sound different to crane ears?

Let's assume, for the sake of simplicity, that cranes do not have a complex language, but only use and understand a few different phonetic utterances as "codes". Such "codes" could mean, for example: Here I am/This is our territory, we belong together (unison call); Come here/Here's something interesting; Take off/Danger on the ground; Danger in the air; Hide/Hide away; We'll fly on; Landing; Good food/Bad food; Sleeping place is safe; you take the brood now, etc.—maybe more, maybe less.

So far, however, I could only distinguish three or four different vocalizations. I did not understand the subtleties, but assumed that there must be a much more differentiated communication than a mere three or four different sounds would suggest. Whether my experiences with the Common Crane would be transferable to the Whooping Crane, I could not say at that time. After all, I had never been able to observe the Whooping Crane in the wild. But I would have the opportunity to do so when I soon traveled to the Texas wintering grounds of Whooping Cranes. Although the DNA of Whooping Cranes and Common Cranes is 99.9% identical, it was clear that the behavior (and even more so the "language") of the two species and individuals would be very different in many aspects. In this respect, I could not expect to transfer my still only rudimentary understanding of Common Crane vocal codes to Whooping Crane communication. And how many different and usable vocalizations of wild Whooping Cranes would I even be able to catch, and in sufficiently clear form, during a one-man expedition of only 10 days in the vast Aransas National Wildlife Refuge?

These were just some of the questions that I asked myself at the beginning of the Whooping Crane project. In the end, it turned out that the communication behaviours of Common Cranes and Whooping Cranes are very different. According to my impression, Whooping Cranes (at least the ones I experienced in captivity) do not communicate with each other in as nearly a differentiated way as the Common Cranes.

Obviously, Common Cranes have undergone an amazing "cultural" evolution over the recent centuries (and possibly only within that time, not before). While, in former times long ago, they used to behave in a clearly territorial manner during the breeding season and in winter, i.e., occupying territories and defending them against each other, these cranes migrate in groups of 20–200 (or many more) individuals and stay in groups for longer periods at resting places and in wintering areas. Whooping Cranes, on the other hand, are invariably territorial—during breeding, migration and in the wintering grounds. The largest migratory flock ever observed for Whooping Cranes included 15 individuals, all juveniles. Normally, the families migrate alone, quite unlike the families of Common Cranes, which seem always to join migratory groups.

The Common Crane, moreover, has even learned, in recent decades, to live in ever smaller patches of wetlands and in closer proximity to people on the edges of villages, while the Whooping Crane apparently has not. What's more, the latter's immediate competitor, the Sandhill Crane, is, in my opinion, even further ahead in regard to cultural and communicative development than the Common Crane, by quite a bit. Sandhill Cranes occasionally even nest where the nest can be seen, their territories are closer together (if the food supply permits) than those of our Eurasian Cranes, and they forage for food in close proximity to human settlements, if necessary. For example, on a field trip in Wisconsin, we observed a Sandhill Crane family with two youngsters filling their bellies in a soybean field, even though that field was right next to a schoolyard. The family arrived just in time for the end of a period of recess, as students disappeared into classrooms for the next 2 hours. The head of ICF's Crane Field Research Project told me that the cranes seem to know the course schedule by now, because they never came during recess (Fig. 4).

Perhaps a lack of flexibility, i.e., the ability to adapt to new environmental conditions, is another important reason for the dramatic situation of

Fig. 4   A dancing Sandhill Crane pair (© Mike Endres).

Whooping Cranes, in addition to past hunting and the loss of wetlands. The example of the Sandhill Cranes speaks to this. Perhaps, despite the loss of wetlands and the fact that they are hunted (both of which were, after all, true of Sandhill Cranes), Whooping Cranes could have survived without the Whooping Crane Recovery Team's elaborate program if only they had become more social and flexible. These speculations are admittedly purely theoretical. But I also wonder if the Whooping Cranes would have evolved a more differentiated "language" if there had been a cultural evolution in them comparable to that of the Common Cranes.[10] After all, there is no differentiated social behavior—whether in humans or animals—without a communication technique suitable for it (which, of course, need not necessarily be vocal). The annual alternation between territorial behaviour in spring and summer on the one hand and social group behaviour in autumn and winter on the other is a constant challenge for Common and Sandhill Cranes, and presumably requires a correspondingly differentiated mode of expression.

A very different question was whether Whooping Cranes in captivity were able to teach their offspring the "language" of their species. However, in captive Whooping Cranes, even these simple communication skills would be

---

[10] Recently, it seems that Whooping Cranes are at least changing their migration behaviour, insofar as they are now migrating in bigger flocks, something never observed before (see chapter "We Fly Off: The Hard Way to the Migration Flight School" and https://wildlife.org/as-whooping-cranes-rebound-flock-sizes-grow/, original study can be found here: https://bit.ly/32qlBf8).

substantially lost, so that Whooping Crane pairs in captive breeding institutions would no longer be able to teach their chick the "language" of their species. According to my considerations, one of our tasks was to find out how the young of wild Whooping Cranes actually learn the expressions appropriate to them.

Even in the event that all efforts to save the Whooping Cranes prove insufficient, for example, because they do not manage to relearn their "language" or because, one day, a leaking oil tanker in the Gulf of Mexico destroys their food base, our work will not have been in vain. The Whooping Crane Recovery Project has already accomplished much more than "only" saving a few rare birds. In Canada, the Wood Buffalo National Park (where Whooping Cranes breed) has been preserved, and in Texas, the Aransas National Wildlife Refuge is the largest of its kind. Without the project, Aransas would have long since been "developed" into a residential, industrial or tourist area. On the migration route from Canada to the wintering area in Texas, numerous wetlands have been secured as potential "stopover" sites for many species of migrating birds. Countless animal and plant species, and thus human livelihood, benefit from these measures. It is not about one species alone: It is about biodiversity, which is only possible in living and sufficiently large biotopes. The exact same thing that is necessary for us humans to live as well.

With the year 1999, my work for the wild cranes changed enormously. Bioacoustics would play an important role in this, because it could serve as an alternative to ringing, which is too risky for animals as rare as the Whooping Crane and the Siberian Crane. It could help us to determine how many cranes live in an area, how long they live, how stable they are in their partnerships, how flexible they are with their territories, and how released cranes behave compared to wild ones. There were new tasks ahead, which took me far away from Duvenstedt Brook. The next chapter had begun.

Research Adventure: Red-Crowned Cranes Eavesdropping at Minus 25 Degrees; Crane Research with Armed Border Guards in the Background

George's favourite cranes—at least the ones he finds the most beautiful—are the Asian Red-Crowned Cranes. And I have to agree with him, because seeing these magnificent birds in the snow on Hokkaido (the northernmost island of Japan) is an impressive experience that you never forget, and one that I owe to George (of course):

On the last evening of my visit to ICF in January 1999, I told him that I was going on a business trip to Korea and Japan 2 weeks later, and that I planned to use this opportunity to visit the Hooded Cranes wintering on Kyushu, the southwesternmost main island of Japan. George was downright aghast: "How can you fly to Japan in the winter, and instead of going north to see the Red-Crowned Cranes, go south to see the Hooded Cranes?" Feeling innocent, I asked what was wrong with that. "In Hokkaido, you will see unique cranes in unique surroundings, an experience you will never forget. I'll call Yulia right away and have her arrange everything, especially the visit with Mrs. Watanabe." I didn't understand what was happening—before I had even agreed, George started making plans for me and calling people. Fortunately, Yulia Momose, ICF's "ambassador" in Japan, didn't pick up, probably because it was in the middle of the night for her.

George calls Yulia Momose his "sister". Once you've met them both, you know why—even if for no other reason than they both have a talent for cabaret and can be hilarious. Yulia is a US-certified internist, but works for the Cranes. Yulia's father was a leading Japanese crane conservationist in his day, and Professor Masatomi,[1] one of the most famous crane professors in the

---

[1] https://www.worldcat.org/identities/lccn-n78013681/, https://www.researchgate.net/profile/Hiroyuki_Masatomi

© The Author(s), under exclusive license to Springer Nature Switzerland AG 2022
B. Wessling, *The Call of the Cranes*, https://doi.org/10.1007/978-3-030-98283-6_10

Fig. 1 An adult Red-Crowned Crane and a young crane, fledged only a few months ago, are flying at a Hokkaido winter day. (© Red-crowned Crane Conservancy Hokkaido)

world, was a constant guest in their house. So, Professor Masatomi became Yulia's "uncle." Eventually, in January, I actually found myself on my way to Hokkaido, to the dismay of my Japanese business partner, with whom I was about to start a joint company. To him, Hokkaido was comparable to the end of the world: "No phones, no infrastructure, no culture." With Yulia, however, I was immediately in the center of the Japanese crane society on Hokkaido. She just knew everyone, and she introduced me to everyone.

The Japanese Red-Crowned cranes, with their pure white plumage and a few black spots on the wings and neck, and the bright red of their crest in the white of the snow, are simply the most beautiful spectacle of nature I have ever seen. My diary from that time still gives me a vivid impression of the freezing cold and the glistening snow in a bright blue sky, the people I met, and, of course, the cranes (Figs. 1 and 2).[2,3]

Truth be told, it wouldn't have taken much for this unforgettable experience to no longer be possible. Red-Crowned cranes still exist in Japan only because, in the early 1950s, people started to feed the last 15 individuals during one of the coldest winters of the last 200 years, and thus saved them from starvation. It was the same in Japan as everywhere else: Centuries ago,

---

[2] cf. https://www.bernhard-wessling.com/Hokkaido-1
[3] cf. https://www.bernhard-wessling.com/Hokkaido-2

Fig. 2 A dancing Red-Crowned Crane pair. (© International Crane Foundation, Ted Thousand)

agriculture was intensified, in this case, rice cultivation. Wetlands on Honshu, the main island, began to be drained on a large scale—destroying the Crane's resting and roosting areas during migration in autumn and spring towards the South and back. At the same time, the time-honored rule that the crane could not be hunted by anyone except the emperor was increasingly ignored.

Thus, the remaining population could no longer overwinter in the south. The crane population was severely decimated and, at times, even considered extinct in Japan. A few birds survived—unnoticed for decades—near Kushiro, where a very large wetland is still located today (it has been a national park for some time, but is still, unfortunately, not secured). In 1923, the cranes were rediscovered, and in 1936, they were declared a "National Monument". A little later, they were again considered extinct, because it had been such a long time since they had been sighted. Their breeding area was not known, and it was assumed that it was outside Japan on the mainland, perhaps in Siberia (it was not until many years later that George discovered it in the wide marshes outside Kushiro). The relationship of the Japanese to the cranes seems ambivalent. Cranes play a major role in Japanese culture, representing eternal life, health, faithfulness—but few Japanese have ever been able to see free-living cranes. Thus, for most Japanese, the birds are a kind of cultural asset, a museum exhibit, but are nevertheless apparently not regarded by society as a whole as a living creature effectively worthy of protection.

For the last surviving Red-Crowned Cranes, there was a hidden, inaccessible refuge, unnoticed by humans. In this vast wetland near Kushiro, there are groundwater and saltwater pools and rivers that normally remain ice-free in winter, allowing just about a dozen of birds to survive and breed successfully for more than 200 years. In 1952, however, all the waters, and even the salty ponds, froze, and the cranes would not have been able to survive the winter. Thus, in their distress, some of them approached the farms, picked up grain and corn near the stables and were discovered by the farmers.

A movement arose among the population to save the cranes: Farmers spread corn; schoolchildren were given crane food by their parents to take to school if it was near a field where cranes had been spotted. The cranes survived the winter.

Since then, corn and fish have been scattered on fields in certain places, which are now regulated by the state and the crane protection agency, and the cranes come and take the food offered. This feeding has enabled a remarkable increase in the crane population. This is a positive development for this crane species, which is so endangered, while the mainland populations are currently still dwindling. Red-Crowned Cranes reproduce naturally on Hokkaido and are able to feed themselves from spring to the beginning of winter, but rely on human help to survive the winter itself.

At the time of my visit, there was a major crane feeding base located at an agricultural school. There, Mrs. Watanabe had been feeding the cranes for decades. After the school closed, she continued the work alone. To avoid having to push the crane feed such a great distance with her cart, she moved the feeding station to her home. That's where I met her.

The feeding station at Mrs. Watanabe's house was one of the places most visited by Japanese people when they wanted to observe and photograph cranes in winter. The gate where the visitors stood is about 100 m away from the cranes. I, however, was allowed to come inside, because Yulia wanted to introduce me to Mrs. Watanabe. We had tea; I explained what George expected of me. "George ..."—a glow passed over the wrinkled face of the old peasant woman. He had left a lasting impression here, too—how could it be otherwise?

George had explained to me what he expected from my visit to Hokkaido: The local cranes had been separated for at least 200 years from the mainland population, which breeds in northeast China and eastern Siberia in the wetlands near the Amur and Ussuri rivers and winters partially in southeast China (on the coast), partially in Korea and partially in southern Japan. While they used to be able to intermingle and interchange, this has not been possible since. George's question was, "Do the two populations have different languages?" Furthermore, he had told me that Professor Masatomi had always been of the opinion that one should be able to characterize and recognize cranes individually by their voice, only he had never been able to develop a suitable method. Now that I had developed one, I should export it to Japan so that banding could be stopped and contactless monitoring would be possible.

So, after tea, I walked through Mrs. Watanabe's kitchen to the back of the house with my recording equipment and camera and stood practically right next to the cranes. Never before or since have I been so close to wild cranes for any length of time. I stayed there (with interruptions) for two full days, sometimes only a few feet from them. My experience with the "Sly Dogs" as I sat in the high seat and they pecked for food below me paled in comparison to a snapshot. After these 2 days, I had a full minidisc with call recordings of many pairs of Japanese Red-Crowned Cranes.

But it didn't stop there. Despite the bitter cold, I stayed outside, because something new was happening all the time. Once, it was a young bird dancing on its own, throwing a colourful piece of plastic around, and I also witnessed arguments between pairs of cranes again. Often, single pairs danced. There was a constant coming and going, as the cranes did not spend the whole day on the snowy meadow on Mrs. Watanabe's property. They flew to other feeding sites,[4] landed in open fields, and foraged in the ice-free rivers. Though dependent on human supplemental feeding, they were by no means domesticated. They knew their helpers, kept their distance from photographers, fled from strangers by the road, continued to forage on their own and slept together away from humans in the river—they dealt with their environment in a very active and observant way.

At the feeding station, the cranes had a reduced flight distance, from the numerous visitors behind the barriers by about 100 meters, from Mrs. Watanabe by less than 10 meters. When I was a guest of Mrs. Watanabe, the cranes also approached me up to a few meters. But when I met cranes in an open area by a road and got out of the car, they flew up even at a distance of more than 100 meters.

My most impressive experience was when I heard a few soft but unusually melodic sounds coming from the other side of a row of bushes where a few cranes were staying away from the feeding meadow. While I didn't hear anything in particular, all of the 100 or so cranes present turned their heads in that direction as if on command.

---

[4] cf. https://www.bernhard-wessling.com/hokkaido-3, Red-Crowned Crane in flight

Then, they began to exchange soft vocalizations among themselves, and moved leisurely toward the row of bushes, with their heads thrust forward with interest. I could not see what was supposed to be there, for Mrs. Watanabe's fodder store blocked my view. Yulia stood next to me, also unable to figure out what was going on. Soon, all the cranes were at the row of bushes, and after a while, some of them began to dance a little, loosely and unexcitedly, while the others gradually went back to the feeding meadow.

A quarter of an hour later, the same game. Again, I could see nothing. When, after another half hour, the same thing surprisingly started again, I climbed onto the feeding boxes. I wanted to know what the cranes were looking at there with such interest! This time, I saw it: a raccoon dog.

It ran along the row of bushes, and then over the far part of the feeding area, after which it turned back. Only now did I notice that, out of sheer fascination, I had automatically recorded on minidisc the sounds that made the cranes turn their heads,[5] but had not taken a photo of the situation. All of the cranes were looking to the left and marching there leisurely, like people in a marketplace where someone shouts, "Here today, for the first time, we are showing a wonderful machine that will allow you to see into the future!"—no one believes it, but everyone goes to see it anyway. I felt as if a crane had called out, "Here is an animal that is not dangerous, but somehow interesting and new," only expressed more simply, in a few quiet vocalizations.

I later told Hiroyuki Masatomi about it, and he replied that he had experienced something like that with "his" cranes, too. He had once spent a night near their roost to observe what happened there at night. It wasn't much. He had noticed that at least some of the cranes were always awake; never were they all fast asleep at the same time. Suddenly, though Masatomi hadn't noticed anything, all of them, even the sleeping cranes, turned their heads toward the shore. They must have heard something, and Masatomi thought it was a soft crane call, signaling something to them. Eventually, they all walked leisurely, as I had also observed, to the shore, contemplated the situation, and then quietly went back to sleep. Masatomi did not find out at that time what the cranes had seen or heard. But he felt that there must have been some signal telling the cranes, "Come here, something interesting, nothing dangerous, but look!" That's exactly how I had felt about it, too.

On another occasion, I heard a sharp hissing sound[6] that even I had no trouble interpreting as a warning, and the cranes even less so: They all immediately stopped feeding or grooming, dancing and playing, bent their necks

---

[5] Call recording "Neugier"(= Curiosity): http://bit.ly/2BTJj2R
[6] Call recording Red-Crowned Crane "Achtung Adler!" (= "Attention Eagle!"): http://bit.ly/2JkqDwY

Fig. 3 A Red-Crowned Crane on Hokkaido defending its food against a white-tailed eagle. (© Ciming Mei)

backwards and pointed their beaks vertically upwards, ready to pounce: A Steller's sea eagle, which is a really huge eagle, came soaring in; it wanted to check if it might be able to hunt down an unwary crane juvenile or an older weak individual (Fig. 3). This reminded me that a German crane observer had once told me that he thought cranes had a special call signal for "eagles", because, at a call, they all looked up, stretched their beaks upwards to ward off an eagle attack, and looked to see where the eagle might be coming from. They reacted completely differently to a fox, much as I had observed with the Red-Crowned Cranes. A few years later, I was also able to hear and record the eagle warning calls of the Common Cranes.[7]

After the recordings I had made at Mrs. Watanabe's house had been evaluated, I was even more convinced that the Red-Crowned Cranes could also be individually characterized on the basis of their calls. Crane research in Japan is very intensive, including the banding of about 20 birds a year in elaborate campaigns. However, the Japanese crane researchers know that they can only capture a fraction of the population with ringing and that they cause disturbance every time they do so.

In addition, the observation is often hindered by natural circumstances: Sometimes the observer is too far away, sometimes the crane is standing in tall grass or reeds so that the rings cannot be seen, sometimes the bird flies away

---

[7] Common Crane warning call "Attention eagle!": http://bit.ly/2WdHPtr

before the telescope is focused. So, it is very difficult to determine a banded crane exactly.

Therefore, the Japanese crane conservationists at that time placed great hope in my bioacoustic method. After all, with "my" cranes, I had succeeded in recording and identifying their calls from a great distance.

I had sent the promising results of the January recordings to Japan, and we were all eagerly anticipating June, when I had to fly to Japan again for business reasons.

For this second stay, I had prepared a CD with the calls recorded so far; I wanted to project them through my megaphone. Because, unlike the Brook, where, in April and May, I could patiently wait for calls at five and four o'clock in the morning, respectively, here, we would wander through different crane territories over the course of the whole day for 4 days, and I wanted to record unison calls and not warning calls. So now, for the first time, I would "call" cranes by megaphone and (hopefully) provoke them into a territorial defense duet. This added to the number of devices I was using for my project.[8] I was aware that the cranes might not respond at all to my call played over megaphone. I had built a very powerful system to make the call audible over long distances, but what would a crane's ear perceive? I hoped that the cranes would accept the CD call as a "real" crane call and respond with their own unison call in defence of their territory.

At first, we only called pairs of cranes that we saw at the same time, so that we could observe their reaction. Each crane pair immediately raised their heads and listened. But the further reactions were very different. The first male crane left the nesting area, approached us briskly and made threatening movements in our direction. So, he must have seen us, but in his opinion, we were probably not cranes, so he searched a bush close to us, obviously wanting to know where the evil intruders actually were.[9] Finding no other crane, he flew back. And when he landed beside his partner, they offered us the unison call we had hoped for, which was captured in its full length and clarity on my minidisc. Later, I got bolder and sounded the megaphone even when we weren't able to see the cranes, which was most of the time. A pair that we only suspected was in the territory came flying in and positioned themselves 250 m away from us. Since it didn't call, I repeated the CD call after a while. This seemed to unsettle the cranes, because they could not see any of their fellow cranes who would be calling, so they flew over us and looked behind the

---

[8] cf. https://www.bernhard-wessling.com/recording-equipment

[9] Vgl. https://www.bernhard-wessling.com/hokkaido-Juni, photographed with a simple first generation digital camera and no telephoto lens.

woods (through which we had come) to see what was going on. After some time, they came back, flew just past us again, landed 250 m away from us and called in duet. Another pair simply flew away, first into their territory and later, after another CD call, over us into a wet meadow on the other side of the road—we did not receive a call.

On another day, we called several times in different territories and had varying success. While a pair leading their chick in our vicinity did not answer, a pair from 1.5 km away on the other side of the lake called back immediately. Here, we were able to "mark" a pair where none had ever been suspected to be before. In all, we found three crane territories that had not previously been known as such.

In a different territory, the male came flying in and uttered very interesting "cooing sounds". I answered with a contact sound that I had on the CD, and the crane answered me, something that I again did not understand. Here, I had the impression—even more so after the analysis—that there was a more differentiated sound expression than the unison call represents.

In the end, each pair, each individual, had behaved differently. These were impressive days, and I sometimes almost forgot my worries about the right buttons, cables and batteries and the constant attacks by ticks in my hair and on my neck. In the evenings, I took a bath in the hot water tub, befuddled by Japanese bath aromas, and removed the ticks.

During the days I spent with Japanese crane conservationists and research-ers on Hokkaido, a nice Japanese-German exchange developed. The bioacous-tic method suited the Japanese very well, as they now had the opportunity to study the birds effectively from a great distance.

I decided to send them an identical version of my recording hardware and offered, from Germany, to evaluate the minidiscs with their recordings from the subsequent years. We had started with an exact localization of the record-ings on maps with a scale of 1:25,000, in anticipation of the next breeding season. My new friends in Japan wanted to record the calls of their cranes and send them to me for evaluation (which, unfortunately, happened only once; the Japanese, actually known for their perseverance, were probably not perse-verant enough as far as crane call recordings were concerned; Yulia recently apologized for the fact that the initiative had not been continued, owing to internal problems; she hopes to be able to use my method again soon).

In the contest over who has the most beautiful cranes, I have come to a compromise with the Japanese and George: The Japanese Red-Crowned Cranes really are the most beautiful, but our Gray Cranes call more beauti-fully. Their call is clearer; that of the Red-Crowned Cranes sounds rougher, more rattling. From the bioacoustician's point of view, on the other hand, the

Japanese calls are more interesting, because they are more richly textured. This suggests a more differentiated "statement" in the call.

My Asian crane observations did not end with my experiences on Hokkaido. During my stay in Baraboo in 1999, I had met "Sunny" Seon Hwan Pea, a doctoral student at Seoul University who had been conducting crane research for many years. And not just anywhere, but in the demilitarized zone (DMZ) on the border between South and North Korea. And that's exactly where I found myself a few months after my Hokkaido adventure. Because once again, George had made plans for me, as well as all of the decisions surrounding those plans. When I first met Sunny at the guesthouse in the evening, he already knew what I was doing, why I was there, and that I had business in Korea, and much more ... So, we just agreed on the details, because by now, I knew George's recommendations as to where I would find things that were interesting, exciting and different.

And so, we drove together closer and closer to the border with North Korea. I felt a little queasy as we approached the checkpoint. We were driving towards the entrance to the civil control zone (CCZ), a zone up to 10 km wide in which civilians are only allowed to stay with special permission. Only then comes the demilitarized zone, 4 km wide, in which—except for a few soldiers on watchtowers—no one is actually allowed to remain for very long, according to the ceasefire agreements of 1953. Sunny, however, had this special permission. He could stay in the CVZ and even in the DMZ, the latter, albeit, only under guard.

For me, the issue there then was whether the cranes on the North Korean border (the Chinese and Russian mainland Red-Crowned Cranes) "spoke" differently than their Japanese relatives.

When we had almost arrived at the checkpoint, Sunny confessed to me that the request for permission for a second person, along with technical equipment, had been made a long time ago, but had not to date been answered, let alone approved. "So, what now?"—"We just go there, because my special permit allows us to take one person, and if they don't check the car, we're lucky. If they do, we'll take another checkpoint and try again." Well, that was a nice outlook. I could already see my expensive equipment being confiscated, or worse: Perhaps we would be checked as we drove out of the border zone with all my recordings, and only then would all my results and apparatus be seized.

My worry turned out to be unfounded. We got through the checkpoint without any problems and spent 3 full days at the border inside the DMZ.

It was a great adventure. As soon as we entered the DMZ, an armed soldier accompanied and guarded us.[10] As we drove the car, he sat in the backseat behind us, his loaded rifle at the ready. While I first felt very uneasy, we later made friends with the soldiers. The Cranes were not abiding by the ceasefire rules, and we wanted to get as close to them as possible. We were interested in finding out if the "language" of the Japanese Hokkaido crane population, which had lived in isolation for centuries, was different from that of the Chinese-Russian mainland population.

We spent the night in a primitive hostel and ate in the open air at a minus 20 °C temperature, a meal at which we were served dog meat, grilled on a glowing fire that barbecued our fronts, while our backs were thoroughly frozen. My research work—recording crane calls—was done under absurd circumstances: I could only obtain reasonable recordings in those brief moments when both sides, North and South Korea, stopped running their droning propaganda loudspeakers for a few minutes, and then only if I was lucky enough that the cranes decided to call during those minutes. To achieve this, I first tried, as in Japan, to get the cranes in the DMZ to respond with calls from Japanese Red-Crowned Cranes amplified via a megaphone. This did not work at all. I tried everything possible, but the Chinese cranes simply did not respond. This was the first indication for me that their "language" was different from that of their Japanese counterparts. In the end, I had no choice but to be there with the cranes early in the dark during the freezing winter nights to record their morning unison calls. After I had the first recordings, I could do control experiments: A mainland crane pair's unison call garnered an immediate response from cranes in the field, while a Japanese one garnered no response at all. Later analyses, which I published, additionally also in a broader publication together with a Russian group,[11] showed that the calls differed in length, but also in many other characteristics. This made it clear that the two populations of Red-Crowned Cranes would no longer be able to communicate with each other, even if they did meet again. It was even possible that a species barrier would develop if the two populations remained separated for a longer period of time.

Along the way, I recorded calls of the white-naped cranes and took assorted photos. It was only when I got home that I discovered something strange in

---

[10] Farewell photo after we had established a relaxed relationship with the soldiers: https://www.bernhard-wessling.com/dmz-soldaten

[11] cf. https://www.researchgate.net/publication/247411615_comparison_RC_Crane_Japan_-_Korea_DMZ; I was later able to make a corresponding publication together with a Russian research group: https://www.researchgate.net/publication/247208727_Between-population_differences_in_duets_of_the_red-crowned_crane

one photo: Next to White-Naped and Red-Crowned Cranes, there was a young, lonely Siberian Crane.[12] When I showed George the photo, he said that he suspected that this crane had probably been separated from its family and would likely not be able to find them again—a bitter loss for this very endangered species!

Before and after these 3 days at one of the most fragile borders and/or ceasefire lines in the world, I was in South Korea on business—what a contrast! Within the DMZ, one almost exclusively saw armed soldiers always at alert, poverty, ice and snow, as well as cranes trying to be heard over bizarre propaganda noise, while in the rest of South Korea, there were lively cities, modern factories, hectic people. Here, loneliness and pitch dark nights, there, day and night traffic noise and, due to constantly flashing advertising on the walls of the buildings, almost daylit nights.

---

[12] cf. https://www.bernhard-wessling.com/dmz-schneekranichjungvogel

The Adventure Continues: Expeditions to the Wild Whooping Cranes

"After another 20 miles, you'll think you've reached the end of the world and maybe have gone too far. But you have to keep going, because it's not until another 15 or 20 miles, looking even more deserted, that you'll reach the entrance to the Aransas National Wildlife Refuge." That's what Tom Stehn, the coordinator of the Whooping Crane Recovery Team, wrote to me in one of his e-mails in late 1999, just before I left for my expedition to the Whooping Cranes' wintering grounds.

I smiled indulgently when I read that, because I had already visited some pretty deserted parts of the world before, for example, in Alaska, Northern Sweden and Iceland. And as long as I was still driving on well-paved two-lane roads, as turned out to be the case in my journey, an "end of the world" feeling didn't really arise. Nevertheless, a nagging uncertainty grew in me about what I should actually do if my rental car broke down, because I hadn't seen another car in a long time. The fact that the 4WD Jeep I had pre-booked was not waiting for me at the airport in "CRP" (= Corpus Christi), but rather, due to an input error by the rental company, was parked at some even more deserted small airport ("CPR" = Casper Natrona County Airport, which even today, 20 years later, sees only five take-offs and landings per day) in Wyoming, was not exactly reassuring. Instead, I had gotten a rental car that was certainly suitable for students traveling back and forth from the dorm to college, but didn't provide me with any confidence at all now in this pitch-dark loneliness, yet what choice did I have? The last 30 miles of my commute to Aransas would have been a poor walk if it had broken down. It was deep night, no light source far or wide. In Germany, you always see a light somewhere nearby, but not here. My car had no interior light, so, to study the map, I had to stop,

B. Wessling, *The Call of the Cranes*, https://doi.org/10.1007/978-3-030-98283-6_11

get out and look at it in the headlights; navigation devices didn't exist back then.

If I had arrived in the afternoon and had seen the last village before the entrance to the Wildlife Refuge in daylight, I might have even wanted to turn back. But I didn't see it until the next day: A completely abandoned campsite was marked "for sale by owner", shops advertising the nearby Whooping Cranes ("Whooper's Nest") were rotten, desolate, and dilapidated, The corrugated iron huts were empty and looked as if they might fly away in the next mild storm.

The few inhabitants the village still seemed to have did not give me the impression of being on this side of the border to the "end of the world", but rather already beyond it, especially since most of the cars were standing around with their hoods up as someone looked less than optimistically into their engine compartments. At least I had already stocked up with plenty of food supply for the 10 days in which I wanted to track the cranes and record their calls, because there was, of course, no shop there, not even one that would have had few options at excessive prices.

I arrived at the refuge in the middle of the night and miraculously found the trailer reserved for me, surrounded by four mobile homes where the volunteers lived. All of this was right at the entrance of the Wildlife Refuge, directly next to the "Headquarters", wooden barracks with offices, where the staff did their work. Tom welcomed me the next morning, and the first thing he did was to show me the maps I had asked for. After all, I wanted to observe cranes somewhere in the landscape, record their voices, and assign them to crane pair territories. Unfortunately, Tom didn't have any topographic maps for me, but rather sketches, certainly somewhat to scale, but what scale were they? In addition, he gave me aerial photos that promised to be of good help. Once a month during the wintering season of the cranes, Tom flew with a single-engine plane and its pilot over all the territories, counted the cranes he detected and noted where he found them. He also tried to identify the few still-banded individuals[1] by having the pilot descend to 20 m and trying to read the rings with binoculars—just listening to the story already had me feeling dizzy.

When I first looked at the maps, I very quickly got the impression that I had taken on far too much: the distances were, compared to Duvenstedt and Hansdorfer Brooks, outrageous. The total area of the Refuge is about 500 km², 50 times larger than both Brooks put together. About half of this is made up

---

[1] In earlier years, Whooping Cranes were banded in the National Park; this practice was later abandoned due to the high costs and risks (for cranes and humans; once, the catchers almost sank in the bog).

by Matagorda Island, where I planned to spend three days. Only 5% of the area in the northeast of the Refuge was open to the public, with a 15-km ring road (this area alone, at about 25 km², is much larger than the majority of all protected areas in Germany and still two and a half times the size of "my" two Brooks combined). Only with a special permit, like the one I had received, could one enter the closed refuge. Not even professional nature photographers received an access permit, because there were enough photos and film sequences with Whooping Cranes, so there was no need for new pictures, and every visitor who did not directly serve the protection of the cranes was an unnecessary disturbance, in the opinion of the refuge management. With only about 180 crane individuals left at the time[2] for the whole species, they didn't want to take the slightest risk. That was why the discussion about my project proposal had taken so long. However, I was finally given a key, which would allow me to move freely in the area for the following 10 days.

The cranes had their territories directly on the coast, and if I interpreted the sketches with which Tom had summarized the territories from his aerial observations correctly, the total coast length on which cranes could be encountered was about 70 km; and even then, I would be able to visit a maximum of 30 pairs (out of the 50 breeding pairs at that time), the others being on islands or peninsulas farther away. I decided to concentrate on these 30 territories in order not to lose too much time with boat and car trips and hikes, because I had to walk most of the mostly quite long distances, and this on pathless terrain. First, given the distances involved, I had to develop a technique for obtaining calls that could be evaluated with knowledge of where the cranes were calling from.

By car, I was able to drive on a service road that led directly from the northeastern end of the area where the entrance and headquarters were located to the southwesternmost end—about 40 km away.

Of course, this service road was not paved; it was a single-lane, very bumpy gravel road. The cars that the refuge staff usually drove there were of an appropriate caliber: four-wheel drive, mighty tires with rough tread—oh, how out of place I felt with my spare rental car that I had received instead of the jeep I had ordered! I had a bad feeling: Traveling a vast area in an unsuitable car that was supposed to take me to the cranes on a gravel road whose stones would possibly burst the slim tires far away from the headquarters. But the cranes did not live close by the road to the right or left of it, but rather 2–5 km away in their territories at the coast of the Gulf of Mexico. How was I going to get recordings of their calls?

---

[2] By now, there are significantly more, as you will read later.

After the pre-ride briefing, Tom showed me the area. We drove into the closed refuge in one of the off-road vehicles, and the route dragged on. We stopped here and there, got on the roof of the car, but didn't see any cranes at first. Suddenly, however, we did: They flew up north of the trail, crossed it, and headed for the coast. My first Whooping Crane observation! As it turned out, these cranes, a pair with a juvenile, had been visiting a waterhole where they usually took fresh water at least once a day.

The waterholes were located on the higher part of the refuge, quite far from the coast, beyond the gravel road. Towards the coast, there was a flat grass and scrub landscape between 2 and 5 km wide, interspersed with ditches, holes and pools. There, the Whooping Cranes wandered around, looking for food, especially cranberries. They preferred even more to be right at the Gulf coast, more precisely, on the "Gulf Intracoastal Waterway" that runs along the coast and on which container ships and oil and chemical tankers sail. There, in the brackish waters of the Gulf, they sought out their favorite food, the blue crab, their essential protein source. Next to the service road were two viewing platforms, not accessible to the public, from which the biologists made some of their observations.

Now, Tom and I stood on one of these platforms and actually saw the first cranes in their territories, at some distance from us, 800 m away as it turned out. One crane was preening its feathers, the other one was poking around in the ground with little apparent motivation. The wind was steady, but not too strong, and I decided to make my first attempts at recording. As I had done 6 months earlier in Japan, I played crane calls by megaphone, this time from captive-bred Whooping Cranes in Baraboo, and was ready to record the response of the territory owners. They raised their heads, paused briefly in their listless activities, but did not respond. I was disappointed. The Red-Crowned Cranes usually responded immediately when I called by megaphone, or they came flying in to check out what was going on in their territory. Why didn't their distant relatives in the US do the same? Perhaps it was because they were in their wintering grounds, and not, as in June on Hokkaido, in their breeding grounds. But I had heard and read that Whooping Cranes also mark territories in winter and presumably defend them against other crane pairs—did they not do so with unison calls?

I asked Tom what he knew about it, and he expressed his rather demoralizing opinion: "I don't think you can provoke and record enough calls with any fairly reasonable effort. We've tried that before—you've already got the tapes, they're practically impossible to evaluate." Yes, those endless empty recordings with occasional guard calls were worthless. "Tom," I replied, "I'm taking a different approach. I don't want to march up to the cranes with the recording equipment running, as you did in your attempts at that time, but I want to remain invisible and wait for their unison calls. I wait for the situation when they *want* to call! I want to record unison calls and not so much the less interesting guard calls."—"You can wait a long time for that," Tom answered, unimpressed, "because they usually don't call." Then why was I there?

We reached the second viewing platform. Again, we saw some cranes in the distance; this time, there were even two pairs and, far in the back, a third, surely more than a kilometer away. We saw the canal even farther away and, in it, a ship—the biggest threat to the world's last naturally reproducing Whooping Crane population. If a tanker or chemical transport ship were to crash, the entire population could be wiped out, because it feeds on the blue crabs in these brackish waters. Compared to this, the risk of erosion from bow waves crashing into the shoreline is relatively harmless, although constant, which is why Tom and the WCRT are so worried. For only where the coastline is artificially secured can the water be prevented from reclaiming more and more land (Fig. 1).

I started broadcasting through the megaphone again, several times at intervals of 30 min each. Each time, I got the same discouraging reaction: One or the other pair raised their heads, but nothing more happened. Maybe I was too far away. No, that couldn't be it. In Japan, I had provoked unison calls even over a distance of up to 2 km! I couldn't make sense of it, and asked Tom if the cranes would at least call in the morning. It was not an uplifting

Fig. 1 A Whooping Crane pair walking through the brackish waters at the coast of the Gulf of Mexico, the male (right, detectable by its size) had caught a crab. (© Ciming Mei)

prospect to have to be on the spot in this vast area at night well before sunrise, but what if it was my only chance? Then I'd have to be out here at the coast line near the cranes' roosts before the crack of dawn. And that despite the fact that I was really craving sleep, because I was drained from jet lag and the previous days' business negotiations. "They don't call in the morning," Tom commented. That would be a glorious excuse to convince myself why I shouldn't even try it in the first place. "How do you know?" I inquired. I was there now, and I was going to be working for ten days in the territory, so I had to be persistent. "I've never heard them call in the morning." A suspicion occurred to me: "Tom, what is 'morning' for you?"—"Well, when I get here, which is about nine o'clock. Of course, you do hear a pair calling sometimes, but that's pretty rare."—"Do you know if they call before sunrise?" I insistently asked. "No, I don't know that, but I don't think they call."—"Have you ever been in this area well before sunrise?"—"No, never." This surprised me: The top leader of the Whooping Crane Conservation Project, in charge of conservation efforts in the U.S. and Canada, apparently had never visited "his" cranes before sunrise.

When I asked if the Whooping Cranes called before sunrise in the breeding area in the spring and summer, Tom told me he didn't know and it would be difficult to find out. "The breeding area is very inaccessible, the only way to get near the nests is by helicopter, and they don't do that at night." How little was known of these cranes! Where should I go during the next days, where in this vast area should I post myself and wait for any calls?

We drove back and discovered an alligator and a poisonous snake on the way. Tom explained to me that I had to be very careful because of dangerous creatures like this. There was also a sign for visitors to the public part of the Refuge warning of killer bees, with the advice that there was no point in running and that it was best to lie on the ground. Oh dear, what prospects were these? What had I gotten myself into? On Hokkaido, I had endured cold winter nights at minus 25 °C, wetlands with mosquitoes and ticks in spring, while in Korea, I was escorted by armed soldiers within sight of the border with North Korea. Here, I was to lose myself in a vast, rough, impassable flatland, repairing flat tires 40 km away from the last post of civilisation and waiting for crane calls and poisonous snakes in the dark late in the night before dawnbreak? And when do alligators or killer bees wake up? I had no idea about these animals and yet was now in a position in which I might meet them directly—or otherwise not be able to record crane calls.

I had imagined that this expedition would be much easier. I had known that different challenges would await me than at home in the Brook (with no idea as to what kind of new challenges), but I had not imagined that it would be so difficult, and even dangerous. Only now did I realize why there were no unison call recordings of wild Whooping Cranes in the form of magnetic tape or modern digital files, in any library of natural sounds in the world, not even at Cornell University, which has the largest bird call archive in the world.

In the afternoon, Tom had appointments and left me alone. I settled into the trailer provided to me by the Refuge. It was a primitive but roomy American-style camping trailer. There was a refrigerator and a small electric stove. The TV didn't work—but I wouldn't have the time or interest to watch TV anyway. In the evening, there was a "cook-out" planned by the volunteers, to which I was invited. I was glad, since it meant that I did not immediately have to dip into my meager canned food supply.

I planned to use the afternoon for a casual walk near the headquarters, on a trail called the Herons Flats Trail, which was open to visitors. It led through a beautiful coastal landscape where freshwater wetlands lay slightly offshore, while seawards, the brackish water formed lagoons and ponds. A variety of waterfowl made their home there. At the end of the trail, where the hiker was expected to either turn around or take the inland return path, a large shallow water lagoon opened up, and in the middle of it stood a pair of Whooping Cranes, sleeping. What a surprise! One of the cranes was banded on its right leg; I guessed it was the female (because it was slightly smaller than the other crane).

What to do? I didn't have my recording equipment with me; after all, I had only wanted to go for a walk and mentally prepare myself for my first expedition day beginning the next morning in the dark. As fast as I could, I ran to the car, drove to my trailer, packed my stuff in my backpack, raced back, marched to the spot where I had seen the cranes about 30 min earlier—and, sure enough, they were gone! I was disappointed, but didn't want to give up yet—maybe they had just moved or flown a little to the side. And lo and behold, only about 200 m from the first spot and not even 150 m from where I was standing, I found them, again sleeping in the middle of the afternoon, their heads tucked into their feathers.[3]

I prepared my devices and played the call that I had also chosen that morning in the non-public part of the Refuge. Their heads came out of the plumage and turned in my direction. There was no apparent excitement; the cranes tucked their heads back into their feathers and slumbered on. Now, hidden behind dense bushes and reeds, I was as close to a pair of Whooping Cranes as I had ever dreamed, and they didn't call!? I thought that maybe they didn't find the call exciting, or was it that it wasn't natural enough? Maybe they

---

[3] cf. https://www.bernhard-wessling.com/herons-flat-pair

didn't feel that their territory was threatened by a crane pair that couldn't even properly call. So, I quietly played all the calls I had on my CD to myself. They were calls from a total of 16 pairs from Baraboo and Patuxent. I discarded the majority of these calls, because they either had too much background noise or they sounded too weak, too indecisive, not self-confident enough, or even as if they had been expressed by sick cranes. In random order, I tried the three calls that, to me, sounded appropriate. I wanted to leave a break of 15–20 min in between each of the three calls in case there was no reaction. And there wasn't any to begin with. At the first unison call, the crane pair didn't even raise their heads. But when I played a call from the pair known as the "Gee-Ooblek", movement could be seen amidst the slumbering pair. They both perked up much more than after previous calls, tightened up, and suddenly stretched their necks and called in unison while I ran my minidisc for the recording. I was electrified, and regained my courage: I had recorded the very first unison call from a pair of wild Whooping Cranes, the world's first![4] I became more confident that I would gradually get to know the Whooping Cranes so well that I would know better what to do to get them to call. For the moment, I decided to play more calls for them. Of these, I can report that neither a Common Crane nor a Red-Crowned Crane unison call generated any impression. In the meantime, the real crane pair had probably come to the conclusion that there couldn't be that many different crane pairs hidden in the reeds, especially not from other species unknown in Texas.

Not bad for the first day. At least I had the first recording, while Tom had rather thought that, with reasonable effort, I would only be able to get guard calls. Of course, I wanted those calls as well. They are certainly characteristic, but less richly structured than unison calls, which can therefore be more clearly assigned to pairs and single individuals. In this respect, I decided to try to provoke unison calls with the help of "Gee-Ooblek" in the following days. George will be very pleased when I tell him this story, I thought to myself. For the crane male "Gee Whiz" had, after all, hatched from the egg of the crane female Tex, which had only been able to be fertilized with semen sent from Patuxent because he, George, had danced with her for years.[5] What a continuation of that former story! Realistically, I would probably be lucky if I managed to pick up a maximum of two pairs a day in the vast, hard-to-reach area, hence a total of 20 pairs at best.

---

[4] Recording of the herons flat pair's unison call: http://bit.ly/32JOuy4

[5] Gee Whiz died in Baraboo in 2021 at the age of 36 years; he had sired 178 cranes; cf. https://www.msn.com/en-us/news/us/first-whooping-crane-hatched-at-baraboo-foundation-dies/ar-BB1epvfj

When I reported this first success to Tom that evening, and mentioned the banded female, he told me that she had not yet been seen this winter—so she had arrived with her mate only today or the day before. What luck!

The cookout in the evening was very pleasant and fun, a nice end to that first day. The volunteers told me a lot about American conservation work. For example, the fact that a wildlife refuge has a higher protection status than a national park. While, in a National Park, visitors are allowed to go almost anywhere and most parks are developed using multiple roads and trails, in Wildlife Refuges, visitors are only allowed to enter relatively small areas. The vast remainder is thus a real retreat for the animal and plant world. A concept that we in Europe do not know at all.

At Aransas National Wildlife Refuge (ANWR), the volunteers took over all kinds of tasks: They ran the "Visitor Center," as well as the switchboard. They cleaned up trash. They repaired, painted, mowed and maintained the trails in the public part of the refuge, gave tours to school groups, stood on the visitor's observation deck and gave explanations. The Refuge couldn't be run without them, at least not in the way it apparently was. In all, the volunteers spent 3–6 months of the year at the Refuge. They received no salary, only the free space to set up their mobile homes. Most of them, but by no means all, were retirees. And all of them were of the opinion that they needed to do something else for society or the earth, as well as for themselves, after or in addition to their previous professional existences.

Full of thoughts, I went to sleep that first evening. Everything was different than I had expected. My mission had even seemed almost futile in the morning, although, with the success in the afternoon, I had become more optimistic. And the work and yearly month-long commitment of the volunteers also gave me a lot to think about. For the next day, I had resolved to get up at a quarter to four.

I got up early the next morning and, after a quick breakfast (with hot tea), drove into the pitch-black refuge. I must admit, the whole situation was pretty spooky. Little lights flashed on and off along the way, the reflection of my car headlights in the eyes of raccoons. Initially, I missed the entrance gate to the closed part of the refuge, but I found it after some searching, unlocked it and drove into the wide, dark loneliness of this wilderness that was completely unknown to me.

I had chosen a place on the map from which I assumed I would be able to hear calls—if the cranes were calling in the morning. Since I couldn't see anything, it was a gamble to find the place from which I wanted to walk towards the coast. I missed it by only about two miles, as I found out later in the daylight. After parking the car, I loaded up my gear and trudged blindly off into the pitch black, trail-less landscape. Only the direction of the wind helped me to determine where the coastline was, a good 2 km away, from proximity of which I wanted to wait for calls.

It was quarter to six; at seven o'clock or so it would be sunrise. At first, I heard nothing but the buzzing of insects. Then, I was startled by a kind of burping and loud smacking right in front of me—was that one of those strange wild boars that are not really wild boars at all and are called "Javelina"? I didn't see anything, especially since I didn't have a flashlight with me—I hadn't expected to be walking around here before dawn, let alone across the wilderness. Around six thirty, the dawn finally broke, and I soon heard the first guard calls nearby; a little later, a Whooping Crane family took off, perhaps 500 m away from me in the marsh. Over the next half hour, I heard five unison calls, at least one of which came from a familiar direction. All were quite distant, but the recordings were good enough for my later analysis. Five recordings, what a great yield, even if they weren't from five different pairs, but maybe two at the most.

Now, I was happy. My suspicion that Whooping Cranes also called in the morning was correct, and they did so even in winter. I could hardly believe that I was now the first person in the world to hear the early morning unison calls of wild Whooping Cranes. This, of course, saddled me with the task of getting up very early every day from then on. But I had taken another step towards understanding this strange and so extremely endangered species.

While the first morning was still mild, the insects buzzing and humming, in the following days, more and more wind came up, and it got cold very quickly. In the early morning, it was often close to zero degrees centigrade, and later even below zero, and that in Texas! I put on two pairs of pants, three shirts, and two sweaters on top of each other, with my thin windbreaker over them, for I had no winter gear with me. Every day now, long before daybreak, I marched anew in the freezing cold into the dark landscape, placed myself at a location from which I assumed that I had several territories in front of me at an angle of about 120°, and waited for crane calls while shivering. I had to set the recording technique in motion with numb, ice-cold fingers.

But it was worth it. I harvested unison calls from at least three different pairs every day, plus a few guard calls. This showed me that the cranes maintained a lively exchange among themselves about territory ownership and whatever else. For me, it was surprising that they also emitted warning calls when they were not disturbed by humans (because, 20–40 kilometers around, there was nobody present except me, and I was invisible). From further observations, I learned over the course of the days that the Whooping Cranes settle territorial disputes predominantly with longer series of warning calls, in contrast to "my" Common Cranes, which prefer to use the unison call for such situations and use the guard calls mostly as an expression against territorial intruders of other species (humans, foxes, wild boars).

Every morning, I would go by car to a different place and stumble off the service road into the dark wilderness. My eyes got used to it, and I slowly got to know the area, so that impassability and belching navel pigs no longer frightened me. Only once did my heart stop—no, rather, it beat violently all the way up to my neck—when I was hissed at in the dark from only half a meter away (in spite of my "heart-attack", I had even recorded that hiss, totally automatically). I think it was a bobcat, because the hissing sounded cat-like, and I had been lucky enough to observe a bobcat a few days earlier in the late afternoon.

Soon, I could navigate the service track in the dark and had memorized tree and shrub silhouettes in the glow of the car lights, aided by the fact that I also drove back in the evening in the dark, because, just before dusk, there were numerous calls to "harvest" when the cranes prepared to roost. Usually, I walked across from the gravel road to the "dike," a small rise perhaps three feet high that had been built in earlier times to protect the pastures from salt water intrusion. From there, I overlooked the marshland and was right on the boundaries of the crane territories.[6] Especially in the morning, in the darkness before dawn, without a flashlight, it was dangerous to stumble through the area, a fact that I was very much aware of, because, if something happened to me, I would be 10, 20, or even 40 kilometers away from the headquarters, and therefore from help. And that was without a radio. Cell phones didn't exist back then.

I had a fixed daily routine. One and a half hours before the start of the day, by between five and half past five, I was at the desired location. Depending on where I wanted to record, I had to allow 1–2 hours for the journey and the walk and get up accordingly early. Depending on the situation or my sense of it, I waited for unison calls or provoked them with my megaphone. After half past seven, the Whooping Cranes didn't call, as they had left the territory to go foraging, so I drove to my base and had a proper breakfast. On 2 days, during the day, I took the boat along the offshore channel with Tom and another staff member. We'd moor at various spots, I'd let "Gee-Ooblek" call via the megaphone and wait for a response. Normally, however, after breakfast, I would spend the time before noon alone in places where I thought I could observe cranes at close range and pick up flight calls and contact sounds. I

---

[6] cf. https://www.bernhard-wessling.com/anwr-deich, back by the water, Whooping Cranes can be seen.

needed these to be able to provide them to the reintroduction project as a communication tool, for the work on the ground and to be amplified with large loudspeakers from the ultralight aircrafts.

After noon, from half past one to half past three, the Whooping Cranes took a siesta, and I did the same. By four o'clock at the latest, I was back at any one of the crane territories with the intention of picking up calls from returning cranes.

One experience in particular was very productive: I was at the coastline, with vast expanses of water, and a chain of peninsulas and islands in front of me. And on the second island, I saw a crane pair. Ah, so there's another territory, I thought, and at about half past five, I decided to ask "Gee-Ooblek" to call out. Unfortunately, the wind was unfavourable and nothing happened. A little later, however, the two cranes took off, crossed the water in low-level flight, came closer and closer, and were preparing to land only about 100 meters away from me. I was well hidden and had an idea: I would play the call again just before they landed, and they would be so surprised that they would respond with their unison call immediately after landing. That's exactly what happened. So, I was able to record pair #17 ("Lobstick"), the male of whom was banded. You will read more about him later.

Similarly, I worked in the afternoons until sunset in many other areas, until late dusk, when I stumbled back through the marsh to the car and then drove to the trailer. By the time I got there, it was dark again. It never occurred to me that I was moving alone within a huge 500 km² area and that this was actually dangerous. The narrow tires on my loaner, which were probably adequate for shopping trips in American suburbia, but not for gravel roads, no longer frightened me, and I just didn't want to think about the possibility of tearing a ligament or perhaps breaking a foot or a leg in the Geest or the Marsh (because the wild landscape there is anything but pedestrian-friendly, especially when you can't see anything in the dark night), of sinking in a wet hole or getting stuck, not to mention alligators, killer bees, and poisonous snakes. In retrospect, though, it seems to me that it was particularly risky because no one was paying attention to when I left the Headquarters or if and when I returned. One thing I always did: I always wore boots against snakes, because my respect for them was well maintained.

I was able to spend three days on Matagorda Island. There, I slept in a former farm building, where, in summer time, school classes were held and students occasionally stayed, led around the island in spring and summer by a couple called the McAllisters to conduct biological studies. The McAllisters lived year-round on the island. They were the island's conservation wardens, employed by the state of Texas, and university lecturers who preferred the absolute solitude of this island to life in the city. They knew everything about the island, and I had already started reading their book on Matagorda. Generally, they left me alone and, in turn, wanted to be left alone, but they were also extremely nice and helpful. Once at night, we had a slightly longer conversation, and it was very stimulating.

Because they were obviously very knowledgeable about the island's overall biology, I queried them as to whether the cranes engaged in unison calls in the morning before sunrise. At first, they said "no", but I was shrewd by now, and, through further inquiry, learned that they too had never visited their cranes before sunrise in one of their territories, not even once—although they had been living alone on the island for years, practically surrounded by cranes in winter. I truly must have been the first person to hear the wild Whooping Cranes' unison calls in the morning.

Matagorda demanded a lot of me. At night, it was around minus five degrees centigrade, which in itself is not a record cold, but in the farm building where I slept, the cracks in the windows were almost as wide as the windows themselves. There were no doors, only open entrances. The wind, which was more like a gale, whistled right through the room, electricity was unavailable (I recharged my batteries during my lunch breaks at the McAllisters', as they had a generator), the candle flame flickered in the storm and constantly threatened to be extinguished.

In my sleeping bag, I wore three layers of clothing: underwear, pajamas, socks and a sweater. The good sleeping bag did its best, and I slept wonderfully. Only my face, and especially the tip of my nose, got freezing cold in the morning. I had spent the occasional night outdoors in this sleeping bag before, but never at sub-zero temperatures. And even then, you had to put on fresh clothes for the night, changing clothes no matter how cold the wind sweeping through the boarded shack was, otherwise you would freeze in the night in your sleeping bag. But I was more than compensated for the inconvenience of my accommodation by the observations that I was able to make in nature during the day. My minidisc filled up with crane calls more and more each day.

On average, I managed to get 35 recordings a day (several calls from each pair), which I sorted in the evening, entered into the maps, logged file by file and saved on my laptop and an external data storage device. Immediately after each call, I would speak into the recorder, indicating the cardinal direction from which it had been heard. Of course, the wind distorted the direction, but it was approximately correct. So, I slowly built up a picture of those territories from which I had already heard calls and those from which I had not yet. I would then post myself there the next morning or evening.

Every day, I added about three new pairs to my collection, one more (or: 50% more!) than I had imagined as the maximum feasible. In the mornings and evenings, I mostly waited for spontaneous calls. During the day, I always worked with the megaphone.

This way, I slowly got to know the cranes better and better. Especially on Matagorda, where I hiked through the marsh, it gradually became clear: If, from the cranes' perspective, a human like me suddenly called like a crane (i.e., if they could see me and then heard a call from my direction), they would, at most, give off a guard call, but I never heard a unison call under these circumstances. As a Whooping Crane pair, you obviously don't call a human to order by a unison call. So, I had to be invisible.

Generally speaking, if you're a Whooping Crane in Aransas National Wildlife Refuge, you don't have to worry about a human approaching you. A brief guard call is sufficient—if needed at all—but then you can keep feeding, or watch what the human is up to. He can't get too close because of the water in between. And if he does, you just fly to the next peninsula. That's how the Whooping Cranes seemed to think, because, unlike the Common Cranes as I knew them, they didn't exhibit the same kind of extreme shyness. If they took the trouble of fleeing at all, they strode away from me very leisurely, continuing their search for food (Fig. 2).

Approaching them within 100 meters was no problem if I managed it slowly. I could sit on the marsh and just watch the cranes. They had no

Fig. 2 Whooping Crane walking in tall grass, looking for food; the photo was not taken on Matagorda, but in a similar surrounding. (© International Crane Foundation, "Whooping Crane in Tall Grass," Crane Media Collective, accessed November 15, 2021, http://gallery.savingcranes.org:8082/items/show/22432)

problem with that as long as there was water between us.[7] If only dry (or wet) land separated us, they retreated slowly, but steadily and purposefully. Never once did a crane pair fly away if they spotted me and I was too close. Whooping Cranes don't get to see humans at all in their very remote breeding range in Canada, and very few in Aransas, so most may only encounter a few humans during migration. Thus, it may be that they did not experience humans as enemies (and their parents had not told them that very few of them are malicious, because those who experienced said maliciousness are dead).

Their perception was very differentiated. They reacted significantly less involuntarily to my megaphone unison call than the Common or the Red-Crowned Cranes. If I was able to hide and then start the call, it was helpful if there was at least one other Whooping Crane in the distance. After realising this, one day in the marsh, I got the idea to prepare a tool for my second stay at the Refuge in 2001: I wanted to use dummy cranes, which I would place on a pole on the dike, visible from a distance.

I got some calls, for example, from the pair No. 7 on an island on the coastal channel, by sneaking up to the shore in a boat, making sure I was not seen, and then calling from the shore. Call provocation worked best when adapted to the situation. Once, I reached the dike on the marsh in the morning before dawn. Ahead of me was an expanse of water that I recognized in the moonlight. From previous observations, I had concluded that the territorial pair No. 6 was roosting near to the shore, so it would normally not be in the water in front of me. It might therefore feel provoked by cranes staying in this water and would be visible by No. 6. My megaphone sounded and I received an immediate response. A reply then sounded from further away to the southwest, which, in turn, provoked the offshore territorial pair into a fresh reply.

When the twilight gave a little more light, I realized why the situation had seemed so real to the territorial pair: In front of me in the water were four obviously juvenile cranes, while the territorial pair stood on the south shore of the same lake with their offspring. The four juvenile cranes had not called, but the pair with the juvenile had responded to me. This pair also showed me that Whooping Cranes do not have fixed roosting habits. In fact, they changed roosting sites almost daily within their large, roughly 2 km² territory (much bigger than the territories that Common Cranes have).

One pair surprised me very much one day: I was once again at the low dike in the darkness of the early morning. As dawn broke, I noticed, to my amazement, that a crane pair was standing very close to me in front of the dike. It had not detected me, and I was able to record a wonderfully clear unison call

---

[7] cf. https://www.bernhard-wessling.com/matagorda-pair, taken with a simple first generation digital camera and no telephoto lens!

160   **B. Wessling**

as they woke up. I waited, laying at my post for quite some time, because I wanted to see why the pair had roosted here on solid ground and not, as usual, in shallow water. Eventually, the two cranes migrated to a nearby black area— and I remembered having seen smoke from a distance around here the day before. Clearly, there had been a fire (it could have been a controlled mainte- nance operation to prevent scrub encroachment), and this pair had drawn an interesting conclusion: "Where there's smoke, there's fire, and after it's gone out and cooled down, there are deliciously grilled and cooked insects and frogs. We'll have to secure those and spend the night close by, so that when the place has cooled down in the morning, we'll be there first before the others." Now, they were feeding at the barbecue bar! This was a strong indication of their ability to plan ahead and the availability of what is called "episodic mem- ory"; more on this in chapter "Can Cranes Think Strategically? More Amazing Observations".

During the days, I usually concentrated on recording calls other than the unison and guard calls of Whooping Cranes; after all, I also needed Sandhill Crane calls for the rehearsal of the ultralight-led migration with cranes of this latter species. There were such cranes in the refuge, also wintering there. They roosted in groups in shallow lagoons, at a respectful distance from the Whooping Cranes, gathered in pairs or, at best, in groups of four juveniles at their various roosting sites. Probably nowhere else is this fundamental differ- ence between the two crane species so directly visible: The Sandhill Cranes change their behavior flexibly. In spring and summer, they are on their own as a pair or a family, defending their territory; in fall and winter, they join together in groups. That's exactly what I saw in Aransas. Whooping Cranes always separate themselves from others as a family, defending their breeding territory in spring and summer, and their feeding and roosting territory in winter (Fig. 3).

The Sandhill Cranes are also more flexible in their foraging. At dawn, the first groups flew out of the refuge to the north to forage in fields outside the sanctuary. By half past seven at the latest, all the Sandhill Cranes had disap- peared. The Whooping Cranes remained in the refuge (this has changed in the meantime as one can find the Whoopers also outside of the refuge, prob- ably because the population has grown).

In the afternoon, the Sandhill Cranes returned; the fly-in began as early as four o'clock and usually ended in late dusk. One of my goals was to collect flight calls that were part of the cranes' vocal communication arsenal, and provide them to "Operation Migration" for training and the migration flight to Florida. I wanted to record these calls at close range, without wind noise, and preferably not from a large number of cranes; ideally, I would record just one pair alone (Fig. 4).

Fig. 3 A Whooping Crane family resting during the migration from North Canada to the wintering grounds in Texas. (© Larry Mattney)

Fig. 4 Sandhill Cranes exchanging arguments with a Whooping Crane. (© Mike Endres)

So, in the afternoon, I looked for places that the Sandhill Cranes might fly over. Having found one, I would sit at the base of a clump of bushes in order to be less conspicuous, because, if you stood out in the open on the plain like a hiker, the cranes always avoided direct overflight, which is something I needed if I was going to capture good recordings of flight calls.

One day, I succeeded. It was sunny and not particularly windy. I was sitting comfortably in my rain pants when I heard a pair approaching from a distance. They were talking to each other. I started the recording. The calls became clearer and clearer, and I finally saw them as they flew in my direction. I had to be very careful not to fidget with excitement and ruin the shot. No seagull cries now, please! Also no ship's horn and no bumblebee buzzing the mike! I'm holding my breath, so you guys please fly right on over me here! And they did. It turned out to be the most beautiful recording I got of Sandhill Cranes. I love it very much and have often played it for other people.[8] Here, I think the cranes are saying, "It's all right, we'll keep flying this way."

Some of my failed attempts also eventually turned out to be productive. On one occasion, a pair of Sandhills came flying toward me while I was standing in plain sight in the marsh. They must have caught sight of me late in the game, became frightened, and communicated that they were now changing flight direction. This call became labelled in the repertoire of my recordings as, "Watch out, danger, we're flying sideways!"

I invested a lot of time in approaching Whooping Cranes during the day so that I could record their very quiet contact sounds, which were then to be used in the actual reintroduction project 2 years later. Already on the first day, I had a plan for this: I wanted to observe which water holes they approached with particular frequency, and then go to the most popular one, hide there and simply wait.

The water holes were located in the higher part of the refuge, north of the service road. The holes themselves had been purposefully excavated, presumably serving to provide fresh water for livestock in times before the Refuge was set up. They must have been 2 meters deep, maybe a meter more in the center. So, if cranes were taking in water there, you couldn't see them until you were close enough to the hole yourself to look in. If you got there after the cranes, you'd scare them right away. So, you had to get there before they did.

On the first try, I had success right away, but differently than I hoped. I had started to build a camouflage, but had forgotten that I first had to bring the microphone to the water and lay out the cable. So, I stepped out of the bushes surrounding the water hole, surveyed the entire hole, and froze: There, just 5

---

[8] Call recording flight calls Sandhill cranes: http://bit.ly/32R74o1

meters in front of me, a 3- to 4-m-long alligator was comfortably laying around. Who knows what would have happened if he had spotted me first. I retreated as quietly as I could, but not before taking a picture.[9] For that day, my courage for waiting for cranes at a waterhole had been used up.

But the next day, I tried it again, in another place. I now first set up the microphone (for which I had brought a tripod), then arranged my camouflage and waited. After not too long, a Whooping Crane pair flew in. But instead of flying directly into the hole, they circled it. Had they had a similarly bad experience as I had the day before, and didn't like being at the water's edge together with an alligator? They flew low over me, recognized me as a new, unfamiliar element, communicated their fright and warning vocally, and turned away. Well, at least that gave me a silent flight warning call of Whooping Cranes on my minidisc, analogous to that of Sandhill Cranes. Nothing else happened for the next few hours; this water hole was off-limits to the pair for now. And no other pair came. Instead, two Sandhill Cranes visited the site, and I was able to record their soft contact sounds. I was thrilled.

The next day, I searched for another hole, because I was still not at ease with the first one. Being an amateur, which I obviously am, I didn't have a camouflage tent with me and had to fall back on the materials that the surroundings of the water hole offered me. But I built a better camouflage, lay down and began to read a book. Simply waiting, not even being able to observe anything (I could see nothing except a narrow section of the water surface), is boring in the long run, and I wasn't in the mood at the time to contemplate myself and my life.

After a while, a wading bird, a yellowleg, began a staccato warning call. It was quite interesting at first; however, I wondered whom he was warning, and about what. I wondered if there was another alligator lying by the water. Or did he not like my microphone? He just kept calling and calling. It was maddening.

That's when some cranes flew in, landing right at the water's edge. The yellowleg stopped its warning calls, and I could hear a crane walking through the water in my headset. But then, the yellowleg started calling again, ruining the recording. When he paused briefly, I heard the soft contact vocalisation of the cranes between themselves through my headset connected to the microphone, which was 8 meters away, right at the water's edge. The recording would not be usable though. There was too much background noise. By now, the wind had picked up, the leaves were rustling, the yellowleg was rampaging. The cranes slowly climbed the steeper wall of the waterhole and became visible to

---

[9] cf. https://www.bernhard-wessling.com/anwr-alligator

me. Through the branches, which effectively camouflaged me, I took a few photos, my first close-ups of Whooping Cranes. They didn't turn out very well with my simple digital camera, but it was nonetheless exciting.

It wasn't until Matagorda that I had any real luck, and this was at a water-hole where I had almost no cover at all. However, it was the only hole I had found that had been approached by cranes, and since I could only stay on the island for 3 days, there was no chance to try different spots. So, I sat next to a bush, more symbolically camouflaged than anything else, since there was nothing I could have used to make a "camouflage tent". At least I was in a small depression, so I couldn't be seen from the waterhole, being behind ground cover. My microphone was again placed directly at the water's edge. First, a Whooping Crane pair with a youngster suddenly appeared behind me in the corner of my eye; they headed for the waterhole flying low. The adult flying ahead saw me, made a sound that I interpreted as "Watch out, danger, we're not landing! Turn back!"—and I captured it on my minidisc.

Only about an hour later, from another direction, visible to me from a distance through the sparse bush, came another pair with a juvenile. They didn't see me, landed in the waterhole and started drinking. For my part, I was drunk with joy, because I could hear the drinking sounds through my headset. I heard them poking through the water, drinking, and I heard their contact sounds. After they drank, they climbed up to the edge of the water hole to where I could see all three of them.[10] Then, they prepared to take off, and I was able to record the sounds that served to coordinate their departure. On the recording, you can hear the wing beats of the cranes taking off and the subsequent communication as they fly—what a sensation, gorgeous, dreamlike![11] Along with the drinking sounds, this was one of the never-expected highlights for me.

Getting so close to the last wild Whooping Cranes was an experience that filled me with deep gratitude. I felt it was a great privilege. I saw alligators, snakes, and, on Matagorda, an Aplomodo Falcon; I watched a Peregrine Falcon strike its prey and then defend it against crows—symphonies of observations and experiences. On the mainland, javelina burped at me in the dark, bobcats startled me out of the darkness, alligators lay only a few feet from me, theoretically ready to eat me. But the Whooping Cranes gave me their calls, letting me share a small part of their lives.

---

[10] cf. https://www.bernhard-wessling.com/anwr-wc-am-wasserloch

[11] Unabridged and unedited recording of the sounds while the cranes were drinking, communicating about taking off and, finally, the take-off itself: http://bit.ly/33zsMgF, The recording also features other ambient sounds (e.g., the calls of a duck).

I was so grateful, moved to tears; I felt an immense, deep happiness. During the days in the refuge, in the solitude, surrounded by this absolute silence, I felt at one with nature. I perceived it with all of my senses, and at that time, I felt particularly clearly that I was part of nature. I felt like Humboldt. In Andrea Wulf's book *The Invention of Nature; The Adventures of Alexander von Humboldt, The Lost Hero of Science*,[12] it says: "In his (Humboldt's) view, knowledge could not be obtained from books alone. To understand the world, a scientist had to be in nature—to feel and experience it—an idea Goethe had grappled with in Faust. ... Humboldt was ... a scientist who wanted to understand nature not only intellectually, but also to experience it with all his senses" (translated by the author from the German edition). As this was my driving principle in my professional chemical research, I also practiced it during my crane research in the Brook, in Japan, and in the DMZ on the border to North Korea. But I felt it particularly intensely there on Matagorda when in such close proximity to the Whooping Cranes. Humboldt would have thought I was a teachable student.

One afternoon, I had made a fairly long hike into the marsh, to the coast, to wait out the dusk and see if I could pick up any more crane pairs from the southern end of the Refuge. Until evening came, I wanted to rest. I laid down comfortably in the overgrown dunes that were there. On the far shore of a large body of water, apparently a lagoon, I spotted a crane pair, noticing shortly afterwards that they had a juvenile. I decided to wait and see what would happen. After a good hour, which I spent reading and watching, I saw a turkey vulture out of the corner of my eye above me—he must have thought I was carrion and came very close. When he was only 2 meters above me, I took his picture.[13] He turned away with disappointment. Another unique nature experience.

At the end of my expedition, I had completed ten very strenuous 16- to 18-hours days, during each of which I had hiked about 20 kilometers. I was very satisfied. My initial concerns had been proven unfounded; in fact, my experiences and results far exceeded my expectations. I had passed all of nature's challenges and had been richly rewarded. I had come within 8 meters of some of the 180 surviving wild Whooping Cranes, had been able to hear their soft contact calls, heard them sipping fresh water and communicating about taking off together, and experienced the sounds of wings beating as they took off. This expedition, successfully accomplished all by myself, in a vast area

---

[12] German edition 2016 by S. Bertelsmann-Verlag, original English edition 2015 by John Murray (publishing house); the citation is a translation of the German text.

[13] cf. https://www.bernhard-wessling.com/anwr-truthahngeier

totally unknown to me, with a task no one before me had ever undertaken, made me happy. When I left the Wildlife Refuge, I was overcome with melancholy; already the very next day, I missed the sunrise[14] after the first crane calls and the setting of the sun at the end of an exhausting but also rich day.[15]

In the 4 weeks after my return, I evaluated all of the recordings I had made in the refuge: I was able to process over 150 unison calls, plus some dozen guard calls, and the evaluation revealed 28 different territorial pairs, eight more than I had set as my maximum achievable goal in my wildest dreams. And last but not least, I had recordings of other Sandhill and Whooping Crane vocalizations that I could use for the planned migration experiments with Sandhill Cranes and later with the first Whooping Cranes to be released. With this, I had convinced the many skeptics in the dozen organizations and institutions that were running Whooping Crane conservation and the reintroduction project in Canada and the USA! I was able to convince Brian Jones, the head of the Whooping Crane Project in Canada, to join me on my second expedition to Aransas. It took place in February 2001. Of the further ten days I spent there, Brian worked with me for three of them. The other 7 days (three of which were again on Matagorda), I worked alone. This time, I was more experienced and even more successful, because I used a new technique: I hid behind the low dike, for example, when I knew there was a pair in their territory not too far ahead of me. I stuck two dummy Whooping Cranes, made at my request, into the ground on the levee.[16] Then, I let my megaphone sound—and, with much more frequency and a greater amount of safety than on my first expedition, I got unison call responses.

In total, I was able to record and evaluate 250 unison calls, and I got to know 32 pairs. I knew 21 of them from the previous expedition. Four pairs had changed their territory. There were six pairs from the previous year that I could not find again. Ten pairs were new to me, at least three of which were in territories that had not existed at the end of 1999. In three cases, I was quite sure that there had been a new mating.

During the second expedition, I was able to provide further proof of the usefulness of bioacoustic analysis. In wintering territory No. 17, during my first expedition, I had identified the pair whose male was banded. Therefore, I knew it was the so-called "Lobstick" male, named after the territory where it bred in the summer in Wood Buffalo National Park in northern Canada. When I began my second expedition in February 2001, Tom Stehn told me

---

[14] cf. https://www.bernhard-wessling.com/anwr-sonnenaufgang-morgennebel

[15] cf. https://www.bernhard-wessling.com/anwr-sonnenuntergang-matagorda

[16] cf. https://www.bernhard-wessling.com/anwr-decoy

that this crane had been seen from the air during an initial count in the breed-ing area, but not during the next count. There was a pair in its territory, but it could be that the many-year-old ring had fallen off, or that the now 22-year-old "Lobstick" male was dead and a new pair, or alternatively, the previous female and a new male, occupied the territory. Tom hoped for clarification from me through my voiceprint analysis. I visited the corresponding territory first, listening to the new recording and the one from the end of 1999 first thing after I returned to my trailer in the evening—they sounded convinc-ingly the same, and Tom was enthusiastic about the result. Later analysis of the acoustic fingerprint confirmed my assignment. The "Lobstick" male con-tinued to be found, even in 2004, during the last recordings—he was 26 years old then, truly a crane Methuselah, and also the most successful male, with the most offspring.

How did these later recordings come about? It started with Brian's interest being aroused, after which he managed to have two biologist positions funded in Canada. I met the two young biologists at a Whooping Crane conference in 2002, where I presented the findings that I had collected so far. In special seminars, I trained them, both theoretically and with practical exercises. I offered to analyse their recordings if they would provide me with files pro-cessed and documented according to my specifications. This worked out excellently.

We monitored Whooping Cranes over a total of nine summers / winters, beginning with my first expedition in late 1999 and ending with the 2004 breeding season. We recorded a total of 918 calls from 236 pairs (if you sim-ply add up the number of pairs per season). Of course, there were not that many Whooping Crane pairs. But, in the analyses, I found many pairs in subsequent years; in total, I was able to identify 80 different pairs, most of which could be found in subsequent or previous years, some not, probably because at least one of the partners was dead.

My analyses allowed for some interesting conclusions:

- Whooping cranes have a high mortality rate, higher than other crane spe-cies. Presumably, this is due to the relatively high degree of inbreeding.
- At least 14% of pairs changed partners, which is a lower number than for Common Cranes. This is probably due to the overall smaller selection of potential mates and to the fact that Whooping Cranes are much less social than Common Cranes. In addition, I may have overlooked new mates

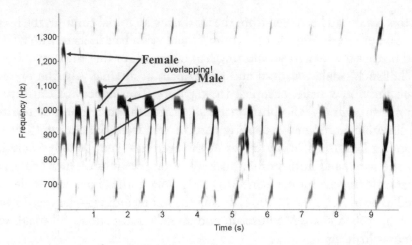

Graph 1 Spectrogram of a Whooping Crane pair unison call recording, showing the partially overlapping frequency range, a characteristic which does practically not occur for Eurasian Cranes

because of the overlap in the acoustic frequency ranges of males and females,[17] a phenomenon that does not occur with Common Cranes (Graph 1).[18]

- 20% of the pairs are found in new territories at some time, another significantly lower number than with "our" cranes, but the territory maps I created already show a remarkable dynamic, a greater amount than all (whooping) crane experts had previously suspected.
- On Matagorda, the pairs are more dynamic than on the mainland, changing territories much more frequently.
- With the help of the "acoustic fingerprint", I could definitely identify eleven more pairs on the mainland in 2003/2004 than were counted by the usual aerial observation by means of monthly flights, and on Matagorda, I had identified three more pairs than the aerial observation had revealed.
- Of particular interest: on Matagorda Island, more than half of the pairs that are neighbours in the breeding area have neighboring wintering territories as well. The same is true on Welders Island and on an area of the ANWR mainland: For the most part, their summer and winter roosts are right next to each other! That means that Whooping Cranes also have something akin to "friendship".

---

[17] cf. https://www.bernhard-wessling.com/schreikranichduett-sonagramm
[18] cf. https://www.bernhard-wessling.com/sonagram-example

Unfortunately, the recordings could not be continued, because the Canadian biologist positions did not receive further funding. Thus, we were no longer able to accompany the genetic analyses of the birds' degree of kinship with similarity analyses of the calls. I had gained the impression that cranes that are related to each other had similar call structures, much in the same way that, in humans, the voices of sons often sound like those of fathers and those of daughters often like those of mothers.

Through my analysis of all calls recorded during these years, 85% of all breeding pairs and 75% of all territorial pairs were ultimately identified. At the end of the study, there were 60 territorial pairs in the then-only wild Whooping Crane population, a total of about 190 individuals. Thereafter, there was an impressive growth of the population to over 500 individuals as of 2021.

We Fly Off: The Hard Way to the Migration Flight School

Today, when some people in Europe read or hear that young Whooping Cranes raised in captivity have been trained to fly behind an ultralight aircraft as a way to learn the migration route to their wintering area, they often say, "Oh yeah, I've seen that on TV before. Other people had done that here before with some other birds, right?"

Yes, there have been other attempts similar to this one in Europe, however, not "before", but rather many years after the successful start of the Whooping Cranes' reintroduction with a UL aircraft-led migration. Probably the most serious European project (started in 2010) with a similar approach involves the Bald Ibis.[1] However, the fact that reintroduction by UL began with Whooping Cranes for the first time in 2001, after years of preparation, is largely unknown in Germany and Europe. Therefore, I would like to describe in the following the winding paths that finally led to the current remarkable (but still unsustainable) success of establishing a new wild Whooping Crane population. This newly reintroduced eastern population moves south in autumn and back north in spring, and it took countless attempts, ideas and contributions from numerous people to make it happen.

It all started when Bill Lishman dreamt as a boy of flying with wild geese one day. Although he first became an artist and specialized in metal sculptures,[2] his dream never left him. He became one of the pioneers of microlight technology, raised Canada geese himself, imprinted them on himself, accustomed

---

[1] cf. https://bit.ly/2MGZxjz, see also https://bit.ly/2SuI4Rx and https://bit.ly/2GBaByz (all in German, last visited Sept. 24, 2021).

[2] http://bit.ly/2LIyd6f (last visited on Sept. 24, 2021).

them to his microlight, and, in 1988, finally became the first person to fly with the geese. In 1993, he demonstrated for the first time, with Sandhill Cranes, that an ultralight aircraft could be used to escort birds south for wintering. His autobiography *Father Goose*[3] became the basis for the world famous film *Fly Away Home* (1996). The fantastic shots that can be admired in this film are not (as some moviegoers today may assume in retrospect) successful computer animations, but were shot directly from Bill Lishman's ultralight aircraft. His approach was exemplary for the preparatory work for the film *Winged Migration* (2001).[4]

But even before Bill became a star, George Archibald took notice of him. He got him involved with the Whooping Crane Recovery Team (WCRT). Cranes raised in captivity show no "migratory restlessness" that would cause them to spontaneously migrate south in the fall and back again in the spring. They simply stay where they are, relying on finding or being delivered food. This raises the question of whether migratory behavior is indeed essentially genetically determined, as is widely read (more on this in chapter "What Can We Learn About Intelligence, Migratory Behavior, the Formation of Culture, Tool Use, and Self-Awareness in Cranes?"). The WCRT was looking for a way that a population of young Whooping Cranes could be released, learn to migrate and become wild. Bill's success with wild geese made him an ideal candidate for the job. The population was to be established in Wisconsin, where a Whooping Crane population had existed long ago but had vanished, and it was to migrate to Florida like that extinct population.

Of course, there were numerous other attempts to teach cranes to migrate. For example, young cranes were brought together with wild crane groups, which were then supposed to inspire them to fly with them. In one attempt with Sandhill Cranes as "foster parents", this had worked well; however, as mentioned earlier, the Whooping Cranes raised among Sandhill Cranes did not want to mate with other Whooping Cranes, but only with Sandhill Cranes, so this attempt was abandoned. In addition, it was unclear whether Whooping Cranes reared in isolation (i.e., not by Sandhill Cranes) would join sandhill crane groups for migration.

---

[3] Bill Lishman, *Vater der Gänse*, German edition Munich: Droemer Knaur 1996. Original English edition by Crown Publishers, 1996; also interesting to read: https://www.thecanadianencyclopedia.ca/en/article/bill-lishman-profile (last visited Sept 24, 2021).

[4] In https://en.wikipedia.org/wiki/Winged_Migration. One can read, in connection with footnote [3] there: "Bill Lishman—He imprinted geese and taught them to follow him in a low-speed ultralight aircraft in a migration path from Canada to Virginia. We owe the idea of the traveling birds to him. Lishman appears in the "Making of" documentary on the DVD release."

In parallel, David Ellis, a scientist from Patuxent, conducted an original experiment. He wanted to accustom his Sandhill Cranes to a truck and guide them with it. There were a number of older findings on this idea that made the possibility seem realistic. After relatively short flight training, he decided to try it. On October 2, 1995, Ellis set out from northern Arizona with a group of cranes in tow that flew at low altitude behind his truck. Despite some weather-related difficulties, they reached their destination, the Buenos Aires National Wildlife Refuge on the Arizona-Mexico border, still a good 600 km away, though not all of the young cranes had flown the entire distance themselves.

High-voltage overhead power lines were the biggest problem. Three times, cranes collided with such lines, with one of the accidents ending fatally. The group simply flew too low. In addition, the route was a confusing zigzag course for the birds, especially since more densely populated areas had to be avoided—one could hardly fly through the main street of a small town! To make matters worse, there were attacks by golden eagles, which didn't cause any losses, but it took 3 days afterwards to collect the scattered cranes again.

The questions as to whether these birds would have survived the winter on their own—the group from this experiment was overwintered in enclosures—and whether they would start the return flight to the summer quarters of their own volition in spring remained unanswered. In any case, the birds did not fly off on their own. They were too tame and felt more comfortable in the enclosure and by the truck than in the wild on such a confusing zigzag return route.

In 1996, Ellis repeated the experiment with 12 young birds, 10 of which arrived at the sanctuary. This time, they were not released together into an enclosure, but individually into different groups of wild Sandhill Cranes. They were quickly integrated there and learned the behaviors needed to survive from their wild relatives without human caregivers. They did not allow humans to come near them after that. At least some of these cranes made the return flight, but their subsequent fate was unknown.

As far as the Whooping Cranes were concerned, the idea was not to let them fly just anywhere with wild cranes, as in the Ellis project, but specifically to the area of origin in Wisconsin in order to establish a new population there. This only seemed possible with Bill Lishman's method: "Fly with the birds."

At George Archibald's suggestion, the WCRT recommended that the first experiments be carried out with Sandhill Cranes. In 1995, they began: Six Sandhill Crane eggs were taken from wild nests in Ontario, the chicks were artificially hatched and they were raised in isolation. Early on, the young cranes became accustomed to the sound of motors. The preliminary result: After long efforts, they actually flew behind the microlight. But the flight to the wintering grounds did not happen, because Bill Lishman could not get a permit from the Canadian government to fly over the border with aircraft and migratory birds. A typical example of the many bureaucratic problems in this project. The first hurdle is to reach internal agreement on which avenues to pursue—no easy task. Then, the WCRT has to convince all of the relevant authorities, and there are many of them.

After the first experiment in 1995, there were further experiments with Sandhill Cranes in the following years, and ultimately a true "Operation Migration". After the cranes had been accustomed to being alone, thus avoiding the proximity of humans but still learning to fly behind the aircraft, a permit to fly over the border was finally obtained—how fortunate that bird migration had evolved before the human bureaucratic authorities could begin their own evolution. On October 24, 1997, Bill was able to fly out with eight cranes. Six birds had not cooperated: They each turned back after a maximum of 5 kilometers, had to be recaptured and were taken by truck to the next stage.

At each stage stop, these six were brought together with the others in the hope that they might join the group after all—unfortunately, mostly in vain. Finally, they were brought directly to Virginia by truck. All in all, it took 22 days to cover 1000 km—a miserable performance from the point of view of migratory birds, but not bad at all for a first attempt of a UL-guided migration.

In the wintering area, the cranes were slowly accustomed to the environment. At night, they stayed in a large enclosure; during the day, they could move freely.

In the spring, there were new problems: Instead of leaving for the north, the six birds, most of which had already not followed the plane in the fall, preferred to fly to schoolyards and to a hospital. So that they could not further disturb the other cranes, which were doing well, they were captured and brought to Patuxent.

The others flew out on March 29, and were observed the following day 650 km farther north, immediately south of Lake Ontario, thus indicating that they were moving much faster than on the outward flight. They had flown in a direct line, but over 150 km west of the outbound route. The lake seemed to have stopped their flight for the time being; they did not fly over the unknown water body, but first oriented themselves westward, skirted the lake, and reached the area where they had fledged 9 months earlier, with an accuracy of about 50 kilometers.

And they did not give up: With strong headwinds, they tried to get closer to the target area, until Bill and a friend arrived in two ultralight aircrafts to guide them "home" on April 11, 1998.

In summer, these cranes became abundantly troublesome, engaging in the previously-mentioned behaviors of flying into human settlements in search of food, landing in kindergartens, marketplaces and even in a prison yard. They were therefore collected and released again in the same place in the autumn. They began a flight along the route just like the year before, this time unaccompanied, but broke it off after a certain distance.

These were the experiences[5] I was told about when I first came into contact with this project at the Whooping Crane Conference in 1998. Through Bill's efforts, it had become clear that it could work in principle: Young cranes reared in isolation could be led south and return of their own accord. However, more care needed to be taken not to habituate them to humans. To this end, I wanted to make a decisive contribution with my crane vocalization research and with my recordings of calls and vocalizations of wild Whooping Cranes, so that human vocalizations would be avoided.

Before we could arrange for the final rehearsal with Sandhill Cranes planned for late summer/fall 2000, I had to process and make usable their calls recorded in winter 1999 on the one hand, and provide suitable hardware for it on the other hand. I had already contacted my former co-worker Frank in advance; he was trained under me as a chemical lab technician at my company, but, after a few years of working there, he had turned his hobby of providing acoustic infrastructure and operating equipment for all kinds of events into an independent business. I provided the necessary information (construction, dimensions of the planes, procedure of raising the chicks and young cranes in isolation, the training, etc.) and decided when, where, how and which crane sounds should be used (starting with the time of the eggs in the incubator). After that, we designed the concept for the hardware together. Frank obtained the necessary individual parts and assembled them: a megaphone with an anchoring facility on one of the rear scaffolding poles of the aircraft; cables; the control unit with which the pilot could select one of the six calls I had chosen. The calls were stored on a microchip, and the chip and control unit were soldered to a circuit board. In addition, we needed a mini-speaker with

---

[5] For the sake of completeness, I want to mention that, also based on Lishman's preliminary work and accompanied by WCRT, another ultralight migration and reintroduction project was carried out by Kent Clegg. However, these attempts—some of them even with Whooping Cranes—were abandoned, in my opinion rightly so, because Lishman's "Operation Migration" simply did a more solid and better founded job. Here is the entire logbook of Operation Migration Foundation, which has been dissolved in the meantime: https://operationmigration.org/site-map.asp; the page is now only accessible via the web archive having last been online and saved on Feb 20, 2020: https://web.archive.org/web/20200220151548/https://operationmigration.org/site-map.asp. However, at that time, not all pages had been saved; the best access to the former logbook seems to be from here: https://web.archive.org/web/20150908023157/https://operationmigration.org/site-map.asp (as saved on Sept 8, 2015).

connections to another microchip. This speaker was to be placed inside the dummy crane used for feeding, from whose beak mealworms could be distributed in front of the chicks with a mechanical pull when the chicks followed the speaker contact sound indicating, "Come here, here is food". The keepers called this device "Robo-Crane".

Together with Bill Lishman and Joe Duff, I mounted the megaphone and the call-control on a microlight in April 2000. Afterwards, I made my first UL flight as a passenger and tested the system in flight.[6] Except for the call dial control, which was not yet simple enough to operate, everything worked correctly. Frank and I then simplified the call choice control so that everything was available in time for the start of the training. During the preliminary phase in Patuxent, non-fledged chicks were to learn to follow the outside (still wingless) microlight in a "playpen". In the next phase, it was the turn of fledging chicks. The rehearsal was again to take place with Sandhill Cranes, for which I had made specific recordings. I had selected a total of six calls: "Watch out!"—"Attention, danger!"—Contact sound—flight call "Everything okay, fly on"—warning call—unison call (territorial defence) (Fig. 1).

Fig. 1 This was not the situation when I recorded Sandhill Cranes unison calls; here, a crane pair is calling while one of the partners is sitting on the nest; I recorded them in winter time in the Aransas Refuge. (© Ted Thousand)

---

[6] cf. https://www.bernhard-wessling.com/ulf-toronto, Bill is the pilot, I am sitting in the back; the megaphone can be seen on the right.

For the start of the real flight training, I was invited to Wisconsin so that I myself could instruct the caregivers, trainers, and pilots on how to operate it. When I arrived at Necedah National Wildlife Refuge, the site that would serve as the future breeding ground for a newly established Whooping Crane population, I was in for a surprise. In all previous attempts, it had taken several weeks until the cranes wanted to fly behind the microlight, and even longer until the whole group did it together. But now: "Everything works perfectly, much better than expected! We could not believe ourselves what we saw: All the cranes were already running and flying in a group behind the plane on the first day,[7] they can't get out of the enclosure fast enough, they are so excited and want to fly along!" I vividly remember myself, when I was in charge of opening the enclosure door, being clipped on the face by the wingtips of a crane rushing out to the UL just before I took a photo (Fig. 2).[8]

When I arrived back at the training camp the next day, there was great wailing, "Two of our young cranes have been kidnapped!" I was confused. "Who is kidnapping cranes here in the wild, and why?"—"Well, a wild pair of

Fig. 2 Two juvenile Sandhill Cranes are eager to take off together with the ultralight airplane which is calling them "follow me" via the megaphone. (© Bernhard Wessling)

---

[7] cf. https://www.bernhard-wessling.com/nnwr-sandhill-follow, second photo taken out of the ultralight aircraft.

[8] cf. https://www.bernhard-wessling.com/nnwr-sandhill

Sandhill Cranes, who apparently don't have any offspring, have been hanging around here all the time, for the entirety of the last few days, and this morning, while our cranes were out, they lured two of them to take off together with the wild adults. They're gone!" I asked, uncomprehendingly, "For days now, you have allowed the wild crane pair to come near our young cranes? Why didn't you use the unison call to defend your territory? That's call number six, I showed you! You're the territory owners, you have to assert yourselves!" In their excitement, the caregivers and pilots hadn't thought of using the territorial defense unison call I recorded to drive away the childless crane pair. "Now what are we going to do?" they asked me. I laughed, "That's easy—we fly off, circle over the kidnapper pair's territory, and let the lure call be heard as we are there."

No sooner said than done. After a quarter of an hour, the two abducted cranes were back with us.

It was a complete success and enormously impressive to me, because I came so often so close to Sandhill Cranes that were not yet, but would soon be wild. I was flying over Necedah in the UL airplane and watched the cranes flying behind—it was like in a dream. My technique worked right the first time, just as I had hoped! No, much better than I had hoped! (Fig. 2)

The later, final rehearsal of the autumn migration to Florida went reasonably smoothly; the cranes returned in the following spring. They behaved quite differently from those in the previous trials. The difference in procedures was almost entirely in the use of my call recordings. In 2001, the first trial was to begin with Whooping Cranes hatched at ICF and Patuxent and then reared there in isolation by humans, having already been put in contact with my recordings in the incubator. This time, of course, it was the sounds of the Whooping Cranes that I had recorded during the two previous expeditions to Aransas. I had selected the following vocal codes for this project and saved them on microchips: Contact sounds—"Follow me" flight call—take-off sound—aggressive defense sound—warning call—unison call. I had also greatly improved the electronics, having included a joystick with which the individual calls could be easily controlled, all installed in a vibration-proof setting under the pilot's seat.

Two months earlier, I even participated in the pre-training of the chicks in Patuxent. Like all the staff working on the project, I wrapped myself in white camouflage robes that were supposed to at least partially mask our human stature. Crucially, however, we never made a single human sound. Those were touching days, because I was deeply impressed to see how a method that I had theoretically devised more than 2 years earlier, and then prepared practically in the wilderness, worked as planned. My skin and stomach tingled with excitement as I watched the little chicks traipsing behind the "Robo-Crane" with my recordings of wild Whooping Crane parental lure sounds, searching for and finding the worms it had dropped.[9]

In September, they fledged. Again, I had been asked to train the trainers and pilots in Wisconsin at Necedah National Wildlife Refuge in the use of the joystick call selection. I arrived in Wisconsin on September 8, 2001. On the ninth and tenth, I participated in the training. This time, the rules were much stricter than in the previous year, but I was still able to get close to the cranes and "communicate" with them using my technique (Fig. 3).[10]

On September 11, I had to take a broken megaphone to Tomah, the nearest small town, to send it from there to Germany for repair. I had driven maybe ten of the planned 30 miles when a special announcement interrupted the beautiful symphonic music on "National Public Radio" (which I tuned into each time I used a rental car when on the road in the US because of the interesting mixture of serious commentary and analyses with classical music): An airplane had just flown into one of the towers of the World Trade Center in New York, and was on fire. I was shocked; within seconds, I grasped the consequences. A terrorist motivation was immediately suspected; there had already been many deaths, but nothing more was clear at first. I stopped in the

Fig. 3 Training flight of juvenile captive (isolation) reared Whooping Cranes with an ultralight airplane (photo taken during the 2009 training phase). (© Joe Duff)

---

[9] cf. https://www.bernhard-wessling.com/patuxent-bodentraining
[10] cf. https://www.bernhard-wessling.com/nnwr-whooper-training

next village and stood at a public telephone on the street (mobile phones were still very bulky and practically unavailable at that time; but there were many open telephone boxes with small enclosures in the USA). I called a business partner in Germany who was in charge of my company's capital investments.

While we were talking about what the consequences of this horrific event would be, he interrupted, describing to me the images that were just broadcast on television, because special broadcasts were running in Germany: "Just now, a second plane has flown into the second tower! Everything is on fire!" Now, it was clear: It was a terrorist attack. Even as we spoke on the phone, nothing was the same as it had been only minutes earlier. I drove on robotically, sent the megaphone to Germany, and drove back.

When I arrived back at the Wildlife Refuge in late afternoon, quite numb, the WCRT team didn't know anything yet. The cranes had undergone a medical examination that day. "Our best crane died of shock in the process!" the staff told me. This was a major setback for our project, of course, but compared to the attacks, which I told them about, that killed thousands, it was practically nothing. The team rushed to a trailer where they had a TV and watched the recent news.

A short time later, it was announced that all flights within the United States were prohibited until further notice—including those using microlight aircraft. The project leaders decided, "We'll fly anyway. Let them spot us at our low altitude if they can, after all, we're only in the air over the Wildlife Refuge."[11] In fact, the training continued as planned. I continued to help out as agreed, but full of worry and anxiety. Phone calls with my family and my company were often interrupted, internet, and thus email connections, were still very slow and especially unreliable at that time, and we had hardly any infrastructure at the Wildlife Refuge anyway. I ended up returning to Germany later than planned, and that only after a number of privately organized, complicated and expensive detours.

In mid-October 2001, the first ultralight-led fall migration began, with three aircrafts flying from Wisconsin to the Chassahowitzka National Wildlife Refuge in Florida.[12] It covered a flight distance of 1218 miles and was successfully completed in early December. Eight juvenile cranes followed the aircraft. Despite some weather-related interruptions, in retrospect, this migration, at 48 days, proved to be the shortest compared to those in the 15 years that followed. Of course, there were all sorts of problems: One of the cranes preferred

---

[11] Take-off for training flight after 9/11: https://www.bernhard-wessling.com/nnwr-start-trainingsflug

[12] For more details, see Joseph W. Duff, "The Operation of an Aircraft-led Migration: Goals, Successes, Challenges 2001 to 2015", in: Whooping Cranes: Biology and Conservation, Elsevier 2018, Chap. 21, p. 393–413.

not to follow the group, but rather wished to fly alone; chief pilot Joe Duff wrote in the logbook, "No. 4 again decided to make his own flight plan, but without coordinating it with the lead pilot." Most of the time, six of the seven Cranes flew together, following the first ultralight. As soon as it became apparent that one or another crane was flying slower, the second UL would take over and fly behind more slowly. Bill Lishman flew at the tail end, supervising the troop. He also reported to the "ground team" the GPS data of the location where crane #4 decided to land so that this bird could be collected. During the training phase, the birds had been in the air together for no more than 27 min; they had to get used to flying for longer periods. After a good 2 weeks, the formation flight also worked better (Fig. 4).

On the way south one night, there was a strong storm. The covered enclosure that was set up every evening for the cranes was blown away, and the cranes flew off in panic in all directions in the middle of the night. During the night, they were collected by the ground team. With the help of the portable megaphones I had provided, the cranes were coaxed out of the various dark patches of forest where they had taken refuge. One crane, however, had crashed into a power line in the dark in its panic and did not survive.

After arriving at the Chassahowitzka National Wildlife Refuge, the six remaining cranes were first housed in an enclosure so that they could get used

Fig. 4 UL aircraft-led migration of isolation reared Whooping Cranes (photo taken during the 2009 migration) (© Joe Duff)

to their new surroundings. They were subsequently moved to a small island in a larger enclosure and then slowly released into the wild.

In mid-December, a crane was killed by a lynx; this animal was later captured alive and released on the mainland, and thus no longer posed a threat. Predatory cats were far from the only problem during the wintering period. It takes a tremendous amount of effort to establish a new Whooping Crane population. All of the cranes had been fitted with radio transmitters so that they could be located as long as the transmitters worked.

On April 12, 2002, the five cranes that survived the winter period began the return flight to Wisconsin on their own. Four of them arrived at Necedah Wildlife Refuge on April 19. Crane #7 (a female) had separated from the group after a short time and had flown on alone. She then spent the entire summer about 75 miles from Necedah with a group of Sandhill Cranes. Occasionally, she was observed with one or another crane from her migratory group, but mostly spent her time alone or in groups of Sandhill Cranes. To my knowledge, none of these "class one" cranes (as they are called by the release teams) bred. Over the years, most of them have fallen victim to predators.

One year later, in the spring of 2003, a total of 21 juvenile cranes returned from wintering, five from the "first class", 16 from the "second class". Two of these were not in Wisconsin, but in Tennessee and Illinois, respectively. So, what these young cranes did after release was anything but predictable.

In subsequent years, one to two dozen young cranes were directed to Florida in this manner each year. Most successfully migrated back to Wisconsin each time, and the next fall, they made their own way to the wintering grounds in Florida in small groups or alone. Young Whooping Cranes were released and led to Florida behind UL aircrafts 16 times in total, from 2001 to 2015. After that, the release procedure was changed; from then on, fledged birds were only brought into the immediate neighbourhood of a flock of Whooping Crane juveniles from previous years that were already experienced in migrating (this had already been tested over the past several years). The new young cranes became quickly integrated into these groups. In total, "Operation Migration" has brought 167 young-of-year cranes to Florida. On average, the number of migration days had been around 21, which was for two reasons: First, the legally allowed daily flight time, along with the air conditions as the warmer middays approached (which also featured an excess of turbulence), resulted in a flight window of only 3–4 h per day; second, the cranes usually did not have enough strength to fly a greater number of hours per day. Therefore, the average flight time per migration was only 32.55 h. That does not mean that the migration would have only taken 21 days, but rather between 48 and 115 days, because, by far, not every day featured weather conditions that would allow the UL airplanes to fly. The wind from the South was a particularly significant reason for the decision as to whether to fly or not.[13]

The new eastern population consisted of over 120 individuals in 2013, but only 101 individuals in 2018. Mortality was very high. From 2001 to 2017, 195 of the southern-bound cranes were found dead or—because they had not been sighted for several years—delisted as dead. The worst year was 2007, when 36 losses alone were reported, 17 of them due to severe weather on the wintering grounds in Florida. About 40% of the deaths were due to predators, particularly alligators. Unfortunately, I had been unable to convince the WCRT team to research the way in which wild Whooping Cranes teach their young to be wary of predators. At least in Aransas, there are plenty of experienced cranes—if we had learned which crane vocalizations indicate danger from, say, bobcats, we could have used them to train the young cranes to be released so as to warn their compatriots. But now, comprehensive research is underway about the predator problem.[14]

According to some scientists, the reason why so many cranes fall prey to these predators is because there are too many predators in Florida and Texas: Their natural enemies, such as panthers and wolves, have all but disappeared. But, in addition, at least 10, if not 20%, of the deaths have been caused by illegal shootings. What people are thinking when they shoot such a rare bird is beyond me. At least, some of the poachers had been caught and punished.

In 2006, on June 22, the first two crane chicks of the new population hatched, the adults being two cranes from the second migration class. In total, as of 2015, 161 nesting pair-years produced 64 chicks, but only nine of them survived.[15] One reason for this are the "black flies", which apparently infect the young chicks with germs, but, above all, torment the adults so much that they either leave the clutch or the hatched chicks. In an elaborate program, many of the breeding crane pairs were thus forced to lay eggs a second time: As soon as the highest density of the "black flies" was detected, the crane eggs were removed from the nests and artificially hatched at Patuxent, at the ICF and at other institutions for later release. Later, UL guided releases were shifted to another wetland area about 60 miles east of the Necedah Wildlife Refuge. This improved the conditions and the reproduction results. It has been calculated that the mortality rate of hatched cranes must not exceed

---

[13] Previously reported details according to Joseph W. Duff, loc.cit.

[14] George Archibald, personal communication.

[15] cf. Joseph W. Duff, loc.cit.

85% for the population to be self-sustaining. With a lower mortality rate, the population would even experience natural growth.

In 2018, the new population consisted of 47 females, 51 males, and only three subadult (i.e., young) cranes. 81 of these 101 individuals were in Wisconsin during the summer, three each in Michigan and Iowa, two in Illinois, and one in Minnesota. The cranes built 23 nests, and, by July 2018 (when I wrote this chapter), several chicks had hatched, six of which were still alive.

That I was able to contribute to this reintroduction project is surely the most valuable outcome of my bioacoustic and sonagraphic research on crane vocal communication. At the same time, this is probably my most valuable research project ever, even compared to my similarly extremely deep and groundbreaking chemical research. The necessarily incomplete account here of the problems and partial successes hopefully makes it clear that my contribution is only one of an incredible number of elements that have come together to make this historic project a success, and that it is far from finished. Many people—employees of government organizations and private foundations, as well as countless volunteers—have invested far more ideas, time, and money than I have.

Overall, the recovery of Whooping Cranes has made notable progress:

- The wild Wood Buffalo/Aransas population comprised 506 individuals in the winter of 2021, almost three times bigger than that from 1999, when I joined the project. It has recently been observed, on multiple occasions, that—in contrast to all the previous years—Whooping Cranes from this population are migrating in bigger groups.[16] Figure 1 in George Archibald's foreword shows one such surprisingly big flock at the Platte River, and recently, a group of about 150 cranes has even been observed in Saskatchewan. This hints at a very healthy development in which this crane species is also learning to transition from a territorial behaviour during breeding season to a more social type of behaviour, at least during migration.
- The newly established Wisconsin migratory population, which is not yet self-sustaining, includes around 80 cranes. In 2021, four chicks had successfully hatched and fledged (comparable numbers to those in previous years). 18 members of this wild population had been wild-hatched in previous years, the remaining bigger number has been released as captive-reared birds. The number of cranes is lower than 3 years before, for the

---

[16] https://wildlife.org/as-whooping-cranes-rebound-flock-sizes-grow/, original study can be found here: https://bit.ly/32qlBf8

reasons listed here.[17,18] It is considered to be a good sign for the development of this population that the pairs that are forming are increasingly dispersing in Wisconsin, looking for and deciding upon suitable breeding sites, i.e., they are behaving more and more independently, just as wild birds do. This was the result of a study run by the International Crane Foundation.[19]

- Another (albeit non-migratory) population was established in Louisiana in 2011. There was a non-migratory population until about the 1940s, but it was steadily declining, and its last representatives were captured after a storm in the late 1940s. The Louisiana population currently consists of about 75 individuals. It was extremely interesting to see reports about a nesting attempt by a Whooping Crane pair in Texas, the first in about 100 years! These cranes came from the new non-migratory Louisiana population.[20]

- Together with some scattered Whooping Cranes, survivors of earlier aborted reintroduction attempts, there were thus 670 Whooping Cranes living free and wild in the USA and Canada in 2021.[21] That's nowhere near enough for the sustainable survival of this species. *Even more shocking was a statement sent out by the International Crane Foundation on Dec 17, 2021: four Whooping Cranes had been illegally shot to death in*

---

[17] https://savingcranes.org/whooping-crane-eastern-population-update-september-2021/. In 2018, this newly introduced wild population numbered a little more than 100 birds, but due to their still high mortality and still too low breeding success (only 5% freshly fledged birds, compared to at least 10% for a sustainable population), it is necessary to replenish the population with young birds to be released. But, as the government had decided to only release parent-reared birds instead of, as practised earlier, isolation-reared birds, in combination with the closing of Patuxent (see next footnote), the necessary number of young birds could not be released.

[18] Unfortunately, the breeding of Whooping Cranes in Patuxent, which had been going on for 51 years, had to be abandoned in 2017 due to budget cuts by the Trump administration. Most of the cranes living in Patuxent were taken to ICF in Wisconsin, with some being taken to the Calgary Zoo (Canada), where captive breeding is also performed. I visited the Calgary Zoo in 1999 and made further call recordings there, but they were not usable. The termination of the great work of the Patuxent Institute by the Trump administration can only be described as a terrible attack on species conservation. This could already be seen after 2018, when the number of birds released was not enough to stabilize the new Wisconsin population, as the pairs transferred from Patuxent to ICF and Calgary have not bred in these new locations, not even three years later in 2021.

[19] H. Thompson, A. Caven, M. Hayes, A. Lacy, "Natal dispersion of Whooping Cranes in the reintroduced Eastern Migratory Population", Ecology and Evolution Vol 11, Issue 18, Sept 2021, https://onlinelibrary.wiley.com/doi/10.1002/ece3.8007, open access

[20] https://wildlife.org/whooping-cranes-nesting-in-texas-once-again/

[21] Figures for 2018: Anne Lacy, ICF, Presentation at the European Crane Conference 2018 in Arjuzanx/France, (Proceedings 9e Conférence Européenne Grue Cendrée 03–07 Dec 2018, Réserve d'Arjuzanx (eds)); for updated 2021 figures, cf. https://savingcranes.org/species-field-guide/whooping-crane/ (last visited Dec 3rd, 2021). This page also shows 138 Whooping Cranes in captivity.

Oklahoma.[22] *This brought the number down to 666 by the end of 2021. This is a particularly tragic crime as so much effort, work and money has been invested in each single wild crane for decades, and yet in just a few seconds, four of them were killed.*

- But anyway, this rebound from just 15 Whooping Cranes in the mid-1940s to over 600 now is a shining example of what conservation and species protection can achieve when government institutions and private initiatives work together, coordinate their efforts, and provide scientific support for their work. The efforts to protect and reintroduce Whooping Cranes and the protection of cranes in the Duvenstedt Brook, Mecklenburg and Brandenburg were all integrated into the protection and regeneration of natural areas. The same applies to the protection of white-tailed eagles, peregrine falcons and golden eagles, as well as to the reintroduction of lynx in Germany.

---

[22] https://savingcranes.org/four-whooping-cranes-shot-in-oklahoma/

What Can We Learn About Intelligence, Migratory Behavior, the Formation of Culture, Tool Use, and Self-Awareness in Cranes?

Are cranes rather stupid or rather intelligent? In other words, do they live essentially instinct-driven lives or can they flexibly and creatively adapt to new situations? The experiences described up to this point indicate a certain intelligence, flexibility and problem-solving ability. Can more be learned? As before, I do not want to approach this question theoretically or by referring to laboratory experiments, but rather I would like to try to answer it with a view into the life of cranes out there in the wild.

I would like to start with the following question: How is it that, in biology communities, the opinion has formed that the migratory behaviour of birds in general, i.e., including cranes, is innate? In the literature, it reads like this [translation by the author]: "Bird migration presupposes an inherited disposition to migrate, with a general direction of migration in southern directions probably being genetically fixed. [...] Moreover, an 'inherited species memory' is likely to predetermine ... migratory disposition."[1] But how do you know that?

This would have required research into how a map, or a specific route or direction, and a time (when to fly) are laid down in deoxyribonucleic acid (DNA) strands, the material basis of heredity, and one could find some publications with results.

Genes control development and life by providing building and construction plans for enzymes. Has anyone ever been able to identify enzymes that are responsible for migratory behavior? Are there enzymes that produce hormones that then trigger the flight instinct, and which enzymes produce the brain cells that are responsible for orientation to the winter quarters and back during migration?

---

[1] Hartwig Prange: Die Welt der Kraniche. Minden: Media Natur Verlag 2016, p. 350f. (This book is available only in German.)

© The Author(s), under exclusive license to Springer Nature Switzerland AG 2022
B. Wessling, *The Call of the Cranes*, https://doi.org/10.1007/978-3-030-98283-6_13

So far, I have not been able to find answers to these questions, which may not mean much. There are many indications that point to genetic bases of migratory behaviour, for example, in blackcaps,[2] in which certain populations have a "southwest migration direction", others a "southeast migration direction", while crossbreeding of these populations results in an almost clean southern direction. Also, the fact that, in some species (for example, the cuckoo), the young birds are able to migrate purposefully alone and without the company of experienced parent birds in the first year (at least, as far as we know) points to hereditary migratory behaviour.

But the corresponding genetic basis has not yet been identified. On the contrary, a group of researchers at the Max Planck Institute for Evolutionary Biology in Plön, Germany, published, in 2017, a broad evaluation of 70 genome studies across all bird families that showed that the groups of genes considered to be candidates for "migration genes" showed no differences between migrating and non-migrating species.[3] So, I wonder why the statement that bird migration is genetically controlled is still repeated so unreservedly in textbooks without comment and in ornithological reference books without any material evidence. One would think that, at least until the final proof of such a statement and/or the refutation of contrary opinions, there would be a scientific obligation to name alternative explanations.

In the following, I would like to put forward two different hypotheses:

An alternative reason for the migratory behavior—at least that of cranes and other species that exhibit similar flexible migration patterns—could be a "culture" that has evolved over hundreds or even thousands of generations. This "culture" would have evolved in parallel with the various changes in climate over the millions of years since birds have existed and been a part of evolution. Such a hypothesis requires us to be willing to grant that birds, in this case cranes, can develop a "culture", i.e., a common habit, that they can constantly change and adapt to conditions and that is passed down through generations.

A second alternative could be found in "epigenetics". Epigenetics refers to heritable changes in the genome caused by environmental influences without triggering a change in the DNA sequence. Ultimately, this concerns the question of how and whether an existing composition of genes becomes active

---

[2] cf. review article: Miriam Liedvogel, Zugvogelgenetik—wie finden Vögel ihren Weg? (Migratory bird genetics—how do birds find their way?), Max Planck Society Yearbook 2016/2017, 1–7, German with English abstract.

[3] Juan S. L. Ramos, Kira E. Delmore, Miriam Liedvogel, Candidate genes for migration do not distinguish migratory and non-migratory birds, J. Com. Physiol. A, Juli 2017, **203**, No. 6–7, p. 383–397. https://link.springer.com/article/10.1007%2Fs00359-017-1184-6

(i.e., in what combination the genes may become active).[4] For example, it has now been proven that trauma can be inherited,[5] as can fear of certain smells,[6] and thus "experiences" in general. All of this is the subject of recent research.

The possible development of a "migration culture" and a possible epigenetic transmission of migration experiences could be combined, and both pathways could furthermore have a regular basis in DNA, i.e., in the genes themselves.

The development of a migratory culture is comparatively easy to imagine, if one is willing to grant birds the ability to learn and solve problems: After the end of the ice age, good breeding grounds became available very slowly, little by little, in the north. In summer, it was easy to live and breed there, to raise chicks and to find food. In winter, however, there was too much snow, and it became cold and inhospitable. So, it was necessary to fly to areas where it was warm and enough food could be found. In the course of a cultural evolution, it could have developed for certain populations that some fly there, and others elsewhere—if all birds tried to winter in the same area, there would be too many there. That's a very simple background, though, for what would certainly be a difficult individual and even more difficult social and cultural learning process. But in the long periods of time that the ice took to retreat, and, even more, that the plants, and thus also insects, required to move in, this process would be quite possible in my estimation, because it could have taken place very slowly over many generations. So goes my conjecture, which is consistent with the observations of recent decades: Cranes are spreading to places where they have not bred for 100 to 200 years, and are now finding suitable breeding territories again.

Can anything more be determined about my hypothesis? Our crane conferences provided opportunities for a fruitful exchange of information. The complexity of the topic becomes clear when looking at the map of European migration routes, and even clearer when looking at maps from 20 to 30 years ago and comparing them with today's maps (Map 1).

---

[4] https://www.zellbiologie.uni-bonn.de/arbeitsgruppen-1/prof.%20herzog/129.-epigenetik-blog-beitrag.pdf

[5] Amy Lehrner, Rachel Yehuda, Cultural trauma and epigenetic inheritance, https://www.cambridge.org/core/journals/development-and-psychopathology/article/cultural-trauma-and-epigenetic-inheritance/8C1FC1DCFF459B4B07F574386627F9DD#, full text here: https://www.researchgate.net/publication/327942788_Cultural_trauma_and_epigenetic_inheritance

[6] https://www.spektrum.de/news/epigenetik-maeusekinder-erben-erfahrungen-der-grosseltern-spektrum-de/1215397 (German); this article refers to the following source (English): Dias BG, Ressler KJ. Nature Neuroscience. 2014 Jan;17(1):89–96. doi: 10.1038/nn.3594., http://www.resslerlab.com/uploads/7/4/9/1/74915911/dias_2014_nature_neuroscience.pdf

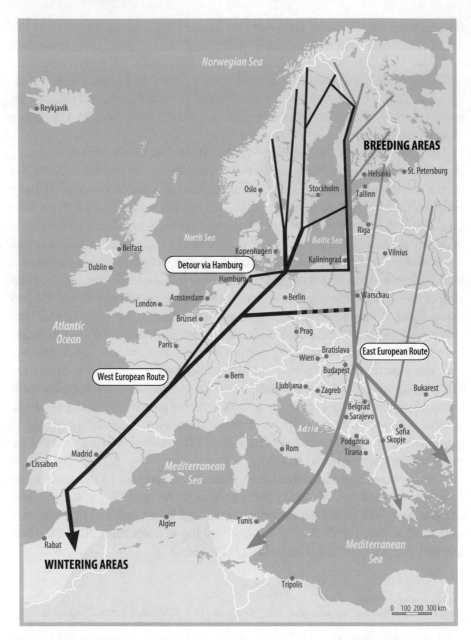

Map 1 Schematic map showing the main migration routes and breeding areas in the 1990s; it is most interesting to see Hamburg as the most western breeding area at that time, since Denmark, Norway, and Lower Saxonia in Germany had no breeding pairs at that time. It also shows the "detour via Hamburg". The wintering areas should also be noted: nothing yet to be seen in southwestern France, no wintering area in the Camargue

The most important resting and gathering places and the main migration routes were clearly visible: a western route via Germany and France to Spain and an eastern route ("Baltic-Hungarian" migration route) to North Africa or Israel and Ethiopia. In addition, there was and still is a migration route running even further east. If the migration routes were genetically determined, those cranes that have "eastern genes" in their cells would have to fly to Africa via Hungary, the others via the western route to Spain. If this were indeed genetically determined, we could even say that the different cranes are of different subspecies, depending on the route.

However, many managers of banding programmes in Sweden and Finland, as well as observers of banded cranes along European migration routes, report strange "irregularities" when collating observations made at German, Hungarian, French and Spanish staging, resting and wintering sites:

- In general, the Swedish cranes used the "west route" to the south, i.e., a route via Ruegen Island or the Mecklenburg Bodden coast, the German low mountain ranges, and then across France to Spain. This would mean that the Swedish cranes had "west route genes". When the bands were read, however, 10% of the Swedish cranes were found to be on the eastern route (ringing is always done on the non-fledged young, which therefore have Swedish parents).
- Of the Finnish cranes, 90% are found on the eastern route, but 10% are observed on the western route, although they should have the "eastern route genes".
- The documents of the European Crane Conference 1996 show that at least a few percent of the banded cranes have been sighted on both the western and eastern routes. The first of these observations was made in 1993, when a crane that had still used the western route in 1991 appeared in Hungary. So, the cranes do change migration routes, at least occasionally. For example, there is evidence that the crane with the number "M 23998" flew the western route in 1992, but the eastern route in 1995 and 1997. But when considering that only an extremely small fraction of cranes is banded (each year, maybe a few dozen out of several hundreds of thousands of cranes), and some of these few banded cranes are already showing such flexible behaviour, we should at least consider it possible that this behaviour is simply common and widespread in cranes.
  Another crane, banded in Finland, turned up in Sweden a summer later—an emigrant. Yet another crane had been proven to winter in Spain via the western route in 1990, 1991 and 1995, but flew via Hungary on the eastern route towards North Africa in 1998.

- In the meantime, several dozen similar observations have been documented, such as that of "M 26215", which flew south over Hungary in autumn 1994 and moved north again over Germany in March 1995.

The latest observations by means of GPS data acquisition of transmittered cranes both confirm this picture and extend it: Cranes are very flexible! This can be seen very well in Map 2.

Such behaviour does not seem to be unique to cranes in the world of birds; it's only that such studies are still so rare with wild birds. Adam Nicholson has collected a lot of information about seabirds. In his book,[7] he tells his readers about the wide individual differences among Northern Fulmars when selecting foraging grounds (where to look for food for their chicks and where to spend the winter on the open ocean) and breeding rocks, and how they cope with changes in the availability of prey. Puffins use similar routes for their long voyages in winter in subsequent years, but every puffin has its own preference. Kittiwakes are even more flexible when searching for fish (made possible by their superior flying performance, but also necessitated by their limited diving capabilities in comparison to puffins, which can dive as deep as 60 meters). Kittiwakes react even more intelligently in regard to variations in catch prospects.

Cory's Shearwaters are even capable of navigating the oceans based on a map they have created by olfactory means.[8] This means that, just as we use a visual mental map on the ground when walking in areas we know well, and as land birds have topological markers in their mental map when flying over land, these seabirds seem to have a map of the ocean with similarly memorizable features as those that we see on the ground or from an airplane when flying over a continent. For us, the vast oceans are absolutely featureless, but not for shearwaters!

All of this does not support a genetically fixed program, but rather the ability for individual learning.

The quite rapid emergence and disappearance of crane roosting and staging sites within one or a few years can also not be explained by genetic programming, but rather by active observation and exchange of experience on the part of the cranes, who freely decide where to gather and where to rest. In Poland, the smaller gathering places have been gradually abandoned, and other places have developed into large gathering locations instead.

---

[7] A. Nicholson, The Seabird's Cry: The Live and Loves of the Planet's Great Ocean Voyagers. Picador 2019.

[8] A. Gagliardo, J. Bried, P. Lambardi, P. Luschi, M. Wikelski, F. Bonadonna, Oceanic navigation in Cory's shearwaters: evidence for a crucial role of olfactory cues for homing after displacement; J. Exp. Biol. **216**, 2798–2805 (2013).

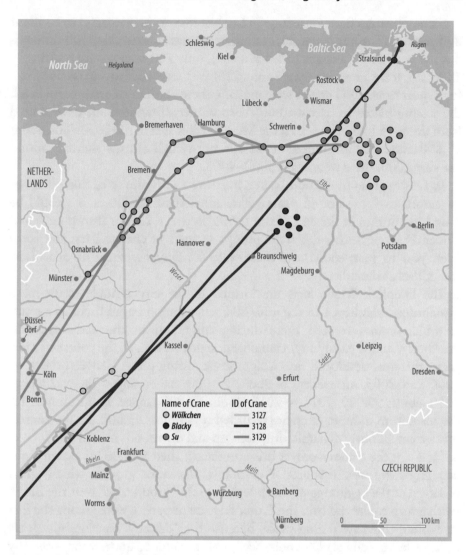

Map 2 This map shows the migration routes of the first three cranes that had been equipped with radio transmitters in Gross Moordorf in 2013/14. The following aspects are the most interesting: The blue route (outward and return) runs almost exactly like the "group journeys" detour via Hamburg almost 20 years earlier, which I described in chapter "Arrival in the Brook After Returning from Wintering Grounds: Alone or in Groups?"—compare Map 1 in chapter "Arrival in the Brook After Returning from Wintering Grounds: Alone or in Groups?" with this one. On the more recent map, the crane also came from a south-westerly direction and turned east, only the change of direction already happened west-south-west of Hamburg and the further route led south past Hamburg. The green flight path leads a little further south from the east to the Diepholzer Moorniederung (Diepholz Moor Flats, north of Bielefeld, where the dots are clustering), then back to the somewhat more "conventional" route where the red flight path runs. On the spring migration, the green flight path goes directly north-east without stopping for a rest in Diepholz

Some examples from Germany: In Luckau (Saxony), there used to be about 200 cranes in autumn, then their numbers increased, to 2000–3500 throughout the year. In Gruenhaus (Niederlausitz), only 700 cranes were roosting in 1985; 15 years later, there were about 2000. There were two maxima when, for a short time, 4000 cranes were roosting there on their migration. All in all, the resting places in Mecklenburg-Vorpommern/Brandenburg have shifted from the coast to the inland. At the Mueritz lakes area in the 1950s/60s, about 15,000 cranes met in autumn; in the '80s, not a single one was sighted; around the year 2000, there were again 3000–4000.

In Linum (Kremmen) and Nauen northwest of Berlin, a resting place has developed over the last 45 years where more than 25,000 cranes could be found in autumn at the end of the 1990s. Between Linum (fish ponds) and Nauen (sewage fields), the crane groups seemed to change back and forth from year to year; sometimes there were considerably more at one place, sometimes at the other.

The Diepholz Moor Flats are located in the city triangle of Bremen-Osnabrueck-Hannover. In the mid-'80s, renaturation began there, as a result of which cranes roosted there during migration in the early/mid-'90s. Previously, no (or hardly any) cranes were sighted there. In the course of a few years, this area developed into a significant resting place: Between the years 2000 and 2005, an average of about 21,000 cranes stayed there; 2006 to 2010, about 50,000; 2011 to 2015, about 75,000 cranes. In autumn 2018, the maximum number of cranes recorded was almost 43,000. As we can see, it is a tremendously dynamic situation, up and down, but overall up.

How did the cranes discover this moorland? There was no migration pathway anywhere near this area! Presumably, the crane groups, which I had observed at the beginning of the '90s during the "group trips" over the Brook (with a turn to the east over the Brook's nature preserve), were among the first to explore this new resting place, because the "detour" via Hamburg leads more or less exactly via Diepholz. Well, that can still be understood—they fly there by chance and find the resting place interesting. But how is it then that, within a few years, there are tens of thousands of them? And sometimes there are 76,000, while in another year, again, "only" 43,000? I cannot imagine this happening without a kind of "beak-to-beak propaganda", without relatively differentiated communication between the cranes, just as much as the "group trips" described in chapter "Arrival in the Brook After Returning from Wintering Grounds: Alone or in Groups?" required such communication.

In the 1980s and early 1990s, we hardly ever observed crane migration over Hamburg, neither from or towards the north nor from/into the east. Since the 2000s, it has become clear that more and more crane groups are migrating over Stormarn/Hamburg, and more recently (around 2015), such migrating flocks have become a familiar sight for us in autumn and spring. In autumn 2018, and likewise in spring and autumn 2019, there has hardly been a day when we have not seen and heard groups of 20 or 70 cranes, or even many hundreds, passing over the Brook on their way to their wintering grounds or back to their breeding grounds. Sometimes, a group descends and rests in the Duvenstedt/Hansdorf Brook or on the large grassland directly south of the Nienwohld Moor.

Here in the Brook and in the neighboring Nienwohld Moor (and, by the way, also at the Schaal Lake and other places in Mecklenburg), I have periodically observed wintering cranes for almost 40 years. I hear and read that birds are seized by the "migratory restlessness", that they cannot in any way resist the migration because of their genes. How is it then that, in mild winters, dozens, sometimes hundreds, even up to around 1000 cranes wintered in Linum/Nauen in the mid-'80s, but in 1987/88, not a single one? In the Brook, too, we occasionally observed one or another wintering pair, which—when heavy frost set in—disappeared for the few frosty days or weeks. I estimate that, in the winter of 2017/18, at least a third of the territorial crane pairs and families wintered in the Brook and nearby Nienwohld Moor. In winter 2018/19, there were only two pairs in the Brook, none in Nienwohld. These pairs disappeared during a short frost period in January; instead, we saw and heard groups of around 40–100 cranes migrating southwest over the Brook almost daily in January (yes, in winter!). Occasionally, they even took a rest here. How can such a thing be possible if the entire migratory behaviour is genetically determined and controlled?

The migration routes have changed considerably over the decades: While the number of cranes on the western route steadily increased tenfold from about 30,000 around 1980 to more than 300,000 in 2011–2015, the number on the central route increased from about 18,000 to 120,000. The much larger increase on the western route compared to the Baltic-Hungarian route can be roughly attributed to three equally relevant effects:

(a) improved crane coverage on all routes,
(b) an increase in the crane population in the areas from which the birds prefer the western route,
(c) a trend towards shifting migration routes from east to west (from Siberia/Kazakhstan to the European eastern route, from there, certain proportions to the western route).[9]

---

[9] Previous information from: H. Prange, op. cit., chapter "Migration behaviour"; but cf. in contrast https://www.nabu.de/tiere-und-pflanzen/voegel/artenschutz/kranich/14589.html. Here, one finds the

These airways are anything but rigid or without alternative. This can be clearly seen in the GPS data of cranes that were radio-tagged in Estonia. A map by Aivar Leito and his colleagues,[10] Estonian crane researchers, shows the migratory routes of cranes, all of which have been tagged not far from each other: They used three different migration routes! Whereas the first few of those cranes used the eastern route exclusively, in later years, only some of the cranes flew on the eastern one, while about 40% used the Baltic-Hungarian (central) one, and as many the western air route. It even happened that cranes using the central route subsequently flew on to Ethiopia—a destination usually approached via the eastern route! Many of the cranes change flight paths, sometimes taking the central, sometimes the western or even sometimes the eastern air route; they change from year to year, or even between the autumn and spring migration. Overall, Estonian cranes take the eastern migration route less often, which was very surprising to the authors.

Another significant example is the female crane "Aino", which had been radio-tagged in Finland.[11] She took off from northern Finland on September 28, 2008, stayed in central Finland for 3 weeks, left southern central Finland on September 18, and arrived in Hungary, in the Hortobágy National Park, a large resting area of up to 150,000 cranes, just 20 days later on September 28. She stayed there for about two months, then flew to Serbia in mid-November, and on to Croatia, after which she crossed the Adriatic and reached Italy. Three days later, she was in Tunisia, having crossed the Mediterranean. She stayed there for about a month, then flew in mid-December, via Algeria (where she stayed for three weeks until January 2009), to Morocco. From there, Aino flew along the northern coast of the Mediterranean to the strait at Gibraltar, reached Spain on January 21, 2009, stayed about 6 weeks in southern Spain, reached the Pyrenees mountain range on March 10, Lorraine (France) on March 24., the Diepholz Moor Flats in Germany on March 27, and Mecklenburg on March 29. On April 8–10, she was in Poland at the

following note, which accommodates my interpretation [translation by B. Wessling]: "Cranes observed in the foothills of the Alps usually start in the Baltic and first fly along the eastern line towards the large resting place in the Hungarian National Park of Hortobágy. Instead of continuing across the Balkans, Turkey, Jordan and Israel like the majority of eastern migrants, the birds turn west. They fly along the northern rim of the Alps and sometimes join the western migrants, but sometimes they also migrate via Italy to North Africa. Why the new migration route has become established is not clear. **Cranes do not have genetically fixed migration routes.**" [emphasis by BW].

[10]A. Leito, I. Ojaste, U. Sellis, "The migration routes of Eurasian Cranes breeding in Estonia", Hirundo 24 (2), 2011, 41–53 (map on p. 48), can be found here: https://www.eoy.ee/hirundo/file_download/65; a further updated map can be found in the presentation by I. Ojaste (co-author A. Leito) at the European Crane Conference 2018 in Arjuzanx France, Proceedings of the 9e Conférence Européene Grue Cendré 03 - 07 déc 2018, Arjuzanx Nature Preserve (eds.).

[11]cf. http://www.satelliittikurjet.fi/aino/aino_gmap.html

Baltic Sea coast, then flew over to Finland, where she again found herself at the starting point of her flight on May 2, 2009. So, this crane flew south on one flight route, visited North African and Spanish wintering grounds, and flew back to Finland on the western route via Diepholz, more westerly than on the main western migration route. Aino spent the winter of 2009/10 in Israel and Egypt.

"Matti" was no less keen on travelling, flying and discovering: He started in the north of Finland on August 16, 2008, spent some time in the more southern parts of the country, reached Poland on October 9, and Hungary on October 13, where he stayed until the end of December. In January to the end of February 2009, he stayed in Serbia, flew back to Hungary on March 8, and arrived in southern Finland on April 23. In 2011, however, Matti moved via Thuringia (May 12, 2011) to France, where he wintered in central France (similar to what he had done in 2010). On April 28, 2012, he arrived back in northern Finland.[12]

Not quite as expansive, but also very interested in other countries, are some cranes hatched in the Czech Republic, which has seen its first breeding successes in the past few years.[13] One of them wintered sometimes in southwestern France (Arjuzanx), sometimes in Spain (Gallocanta). Another flew first to Poland (which is north of its breeding grounds!) in the winter of 2017, then back to the breeding area in the Czech Republic, then again to Poland, then to Brandenburg, and from there to France. Another crane hatched in the Czech Republic and then banded flew *north*west (!) from the breeding area in March 2018 and stayed in the Diepholz Moor Flats. How did it know this place? If it flew with its parents, why did they fly there from the Czech Republic? How did the parent cranes know of this place?

In southwest France, near the Atlantic coast at Arjuzanx, a completely new wintering site was created towards the end of the 1980s in a former phosphate mining area after renaturation. Within only 10 years, it had developed from a resting place for a handful of cranes to an important wintering area for about 20,000 cranes; in 2001, there were once even more than 30,000. Further smaller wintering sites developed in France in Lorraine (500 cranes) and Champagne-Ardenne (about 6000 cranes) (Fig. 1).

The use of migration routes is as flexible as the use of gathering and resting places. Although there are preferred corridors, these have differentiated significantly in comparison over the last two to three decades, as Map 3 shows.

---

[12] cf. http://www.satelliittikurjet.fi/matti/matti_gmap_2008.html

[13] Petr Lumpe, Marketa Tikhackova, presentation at the European Crane Conference 2018 in Arjuzanx/France, Proceedings of the 9e Conférence Européene Grue Cendré 03 - 07 déc 2018, Arjuzanx Nature Preserve (eds.)

Fig. 1 Six Eurasian Cranes flying into the roosting sites in the Arjuzanx wetlands where, only since recently, about 20,000 cranes have been wintering. (© Bernhard Wessling)

These examples show that my suggestion that we regard the migratory and resting behaviour of cranes, at least hypothetically, as a "culture", i.e., a common habit learned through experience, socially communicated and passed on within one generation (and then across generations), instead of assuming genetic bases, cannot be so easily dismissed.[14] Just to show you that this is not a unique feat of particularly flexible European cranes: There are six crane species in India, one of which is the Eurasian Grey Crane, which winters there—coming from Siberia. 20 years ago, the Indira Gandhi Canal was completed in India. Immediately after, a few cranes emerged to winter in the newly created wetlands. Previously, they had wintered some 500 km farther south.

Only a few years later, 5000 cranes were wintering in the artificially irrigated areas. This saved them 1000 km of energy-sapping flight. It is interesting that a few cranes will first "try out" the new wintering site, so to speak, and then, over the course of the following years, more and more animals will join them. Do the other cranes learn from the first, more daring ones what the wintering at the new place is like? In any case, new common habits are formed—this could be called "culture".

---

[14] Some of the examples described above have been disseminated at conferences, many others can be found in: H. Prange: Die Welt der Kraniche (The World of Cranes), op. Cit. (book only available in German).

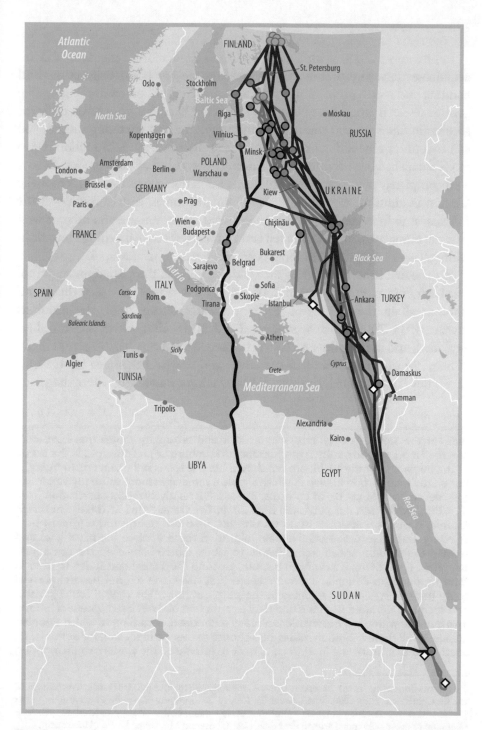

Map 3 In contrast to Map 1 from the 1990s, this map of the main migration routes as of around 2010 shows a clear widening, especially on the western route (yellow) and a distinct central migration route (green); the eastern migration route is shown here in

(continued)

In December 2018, at the European Crane Conference, I met Guo YuMin, a Chinese university professor from Beijing. He has been focusing on bird banding and tagging since 2014, with a particular focus on cranes.[15] So far, he and his research group have been able to track 199 cranes from seven different species (in China, if you count breeding, migrating and wintering species, you can encounter nine different species). By the end of 2018, 1.25 million pieces of GPS data had been accumulated. Among other things, Professor Guo discovered completely new, previously unknown migration routes for the Demoiselle Cranes: In autumn, they migrate from Mongolia on an eastern route over the Himalayas to India, while in spring, they avoid the mountains and migrate west past the Himalayas, first north, then east to their respective breeding grounds in Mongolia—a culturally traditional behavior! Please look at Map 4.

For the first time, a wintering area is developing in the Camargue region of southern France. From 1999 to 2005, only a few cranes were seen there, and none at all in the winters of 2001 and 2002. From 2008, the numbers increased: In 2009, there were over 2000 birds, after which the numbers exploded, and in the winter of 2017/18, there were about 16,000 cranes there. Where did they come from so suddenly, and how did this new habit ("culture") develop so quickly? Where did they winter previously? Over the years,

---

←

red. For the sake of clarity, cross-connections and secondary routes that have only emerged more recently, such as the "detour via Hamburg" that I observed in the Brook at the beginning of the 1990s, are not shown here. It too is still flown less frequently, but cf. the map section in Map 2, which shows a somewhat more southerly variant of the "detour" via Hamburg. Of particular interest is a relatively recent observation from late 2016, which was not published until 2019: The crane "Ahja 5" (black line) from Estonia, having an attached radio transmitter, flew the central route (green) over Hungary, but then turned off the main migration route (look at the black line) and crossed the Mediterranean from Albania to Libya, where it reached the coast near Benghazi. From there, it flew further south, crossing the Sahara and Sudan to its wintering grounds in Ethiopia. It is remarkable that this crane did not use the eastern route (red), as was always assumed in the past, but rather the central route (green), and then also deviated from this and flew a route that had not been observed before. This is in line with other observations already mentioned, according to which Estonian cranes use all three migration routes, also choose to change them and, for example, fly south on the central route in winter and back to Estonia on the western route in spring

---

[15] Guo YuMin, lecture at the European Crane Conference 2018 in Arjuzanx/France. Proceedings of the 9e Conférence Européene Grue Cendré 03–07 déc 2018, Arjuzanx Nature Preserve (eds.); later publication as a peer-reviewed paper: "Time and energy minimization strategy codetermine the loop migration of demoiselle cranes around the Himalayas" Chunrong Mi, Xinhai Li, Falk Huettmann, Oleg Goroshko, Yumin Guo, Integrative Zoology, January 21st, 2022; full papaer available here: https://www.researchgate.net/publication/358009678_Time_and_energy_minimization_strategy_codetermine_the_loop_migration_of_demoiselle_cranes_around_the_Himalayas

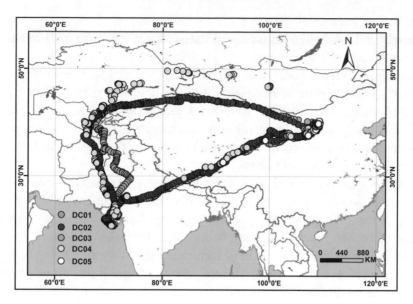

Map 4 Here, you can see the migration route of several Demoiselle Cranes (DC01 to 05), which were equipped with radio transmitters by Professor Guo's team (China). The GPS data show the crossing of the Himalayas in the autumn migration, while the cranes flew west around the Himalayas in spring. So, they did not use the same route as on the outward flight. It is assumed that this has developed due to usual wind directions which are favouring these routes. (© Guo Yumin)

24 banded cranes have been identified, of which 19 came from Finland, four from Estonia and one from Germany.[16]

The degree to which cranes are flexible with regard to migration routes, choice of breeding sites and, finally, their choice of mates is also shown in the DNA: Cranes from Eastern, Western and Northern Europe show practically no genetic differences, and the genetic structure testifies to mixing that is both regular and very active[17]—cranes from Russia migrate to Finland and find mates there, cranes from Finland breed in Germany, and cranes fledged in Germany find mates and breeding territories in Sweden. All of this is evidence of long exploratory flights and a lot of communication, especially certain at resting sites and wintering areas. And it is a serious argument in favour of my hypothesis that migratory behaviour may be a culture that is acquired and passed on over generations.

---

[16] A. Salvi, presentation at the European Crane Conference 2018 in Arjuzanx/France, Proceedings of the 9e Conférence Européene Grue Cendré 03 - 07 déc 2018, Arjuzanx Nature Preserve (eds.)

[17] M. Haase, presentation at the European Crane Conference 2018 in Arjuzanx/France, Proceedings of the 9e Conférence Européene Grue Cendré 03 - 07 déc 2018, Arjuzanx Nature Preserve (eds.).

Biologists have a well-known test for whether animals have an awareness of themselves: They hold a mirror up to them. Great apes such as chimpanzees can recognize their own mirror image: When scientists painted a spot on a chimpanzee's forehead and then held the mirror up to it, the ape wiped the spot off. In my opinion, however, this experiment does not allow for the reverse conclusion that animals that do not recognize themselves in the mirror (but, like budgies, for example, always peck against it, likely believing that they are seeing a competitor or potential partner in their reflection) basically have no awareness of themselves.

"(Self-)Consciousness" is a very complicated phenomenon, one that we mostly still do not understand. We can only look at it superficially here. First of all, we should realize that "consciousness" can have different qualities. It is a kind of consciousness if a living being can distinguish whether it is either perceiving something about itself or something about another living being (of the same species, for example). To be able to distinguish this already testifies to a certain level of "self-awareness".

This distinction between "me" and "someone else" is essential for survival in many situations. For example, every living creature must react differently to a situation in which it is being hunted itself than it would to one in which another member of its herd or group is being hunted. A wolf will not look around anxiously in anticipation of an approaching territorial owner if he suddenly smells his own territorial mark. He recognizes his mark and knows that it is his own scent. This is a kind of consciousness, but a consciousness of a different quality than when a living being recognizes itself as "I" and thinks about itself. We have no information whatsoever about whether and to what degree living beings other than humans are able to do this.

I would like to mention another aspect: "Consciousness" requires, as one of many criteria, a certain "image" of oneself. Such an image may be predominantly visual in the case of optically oriented creatures; but in the case of animals that orient themselves mainly by their noses, it may be shaped by the sense of smell. For example, I would like to ask a biologist who does not grant a dog or wolf any consciousness of itself whether this animal really does not recognize itself when it marks its territory with its urine.

Is he aggressive towards his own scent-marking? Does he defend his territory against himself? No, of course not, he recognizes himself! So, he knows something about himself, namely, at the very least, how he smells in contrast to others. And maybe he even has a visual image of himself, only it doesn't match the image that a mirror shows him (maybe because his reflection doesn't smell "like dog", especially not like himself).

Woodpeckers have the habit of hiding from passing pedestrians behind a tree in such a way that they can just see the passers-by, but the humans can hardly see them (or only their heads). For years, I have enjoyed engaging in the following experiment when I encounter a great spotted woodpecker while walking: I change my position back and forth so that I would be able to see the woodpecker from the other side of the tree. But he always changes his position in relation to me so that his body is hidden behind the tree, with his eyes remaining fixed on me. So, the woodpecker must have some sort of mental image of himself that shows him what it might look like being obscured by the tree. At any rate, I know of no other interpretation.

The cranes' "image" of themselves became clearer to me through an observation related to bioacoustics. For his recordings of crane calls, Eberhard Henne played a unison call with a tape recorder very close to the crane territory, whereupon the territorial pair answered. In order to record this answering call, he had to react quickly, because the territorial pair often immediately fell into the call of the "tape pair" and answered vigorously in unison. This frequently frustrated his efforts to get a good recording, because Eberhard could not stop his old mechanical tape recorder and start the other tape recorder fast enough to record the initial response.

However, there was a single instance when he had to play the tape call over and over again until the cranes responded, and that was with the crane pair whose unison call recording he had just played. Eberhard wondered if the cranes had recognized themselves. Perhaps, I said, they didn't respond because they wondered, "How can we be out there calling when we ourselves are in here?" I repeated this experiment myself a few times, sending a megaphone unison call to a crane pair I knew using a recording of their own call—the pair didn't answer, but I was sure that they were in the territory, because I had either heard or seen them shortly before.

In this case, the cranes' "image" of themselves would be acoustic. Perhaps cranes recognize their own voices even better than we humans recognize ours. Those who hear their own voices for the first time on tape or on the radio hardly recognize themselves. This is because our head is a resonating body that carries a different version of our own voice to the ear than the sound waves in the air carry. This is not the case with cranes.

So, our own acoustic image of ourselves is imperfect, and our olfactory one certainly even more so. If wolves were to conduct consciousness experiments on us, they would not hold a mirror up to us, but an olfactory sample from our own armpit and some from other peoples' armpits and ask us to determine which one is ours. I am afraid we would fail the test and be considered an animal species without consciousness of itself. Cranes would perhaps play us our own voice from a tape recorder, and if we had never heard ourselves through a loudspeaker, we would fail that test too. Perhaps our ideas of what consciousness is are simply different from those developed by other creatures.

However, cranes seem not only to have an acoustic idea of themselves, but a figurative one as well. It is well known among crane watchers that some of our Grey Cranes (not all!) gradually acquire noticeably browner back feathers during the breeding season. It was initially thought that this was due to a change in colour pigments in the feathers, perhaps due to exposure to sunlight.

It is now known that the cranes look for places on the ground where they can pick up brown mud containing iron oxide pigments with their beaks. They actively rub their back feathers with it. This makes them more and more brown over time. I had observed this myself once, but could not make sense of it until I spoke to Bert Brueshaber, one of the discoverers of this phenomenon, and he pointed me to his observations and his publication.[18]

In Sweden, Brueshaber was able to observe that cranes actively turned rusty to dark brown in the manner described above, and always during the period when nature changed its colours, from mid-April to early May. The various crane pairs behaved differently: In one pair, it was mainly the female that turned a determined rusty brown, in another pair, it was both partners. Brueshaber writes: "The female crane walked single-mindedly from the breeding site towards a point at the edge of the moor at the foot of a mountain. There, from a depth of about 13–15 centimeters, she retrieved a rusty-brown, slimy mass, placed it on her shoulder area with her beak, and stroked it out from her shoulder over her back with her beak, head, and neck. This took 8–10 minutes. Now, the crane posed itself so as to dry erect with its wings toward the sun."[19] [translation by the author].

After 5 minutes, the female went back to the nest and exchanged places with the male. The male now went to the same place, but not quite as determinedly, and also colored itself, but "more slowly and not so thoroughly". Did the male not really feel like it? In another pair, Brueshaber observed that the two were colouring at the same time. Still another pair, which did not have any suitable moor soils in its territory, flew into the immediate vicinity for colouring.

The background to this behaviour is probably that cranes often breed in areas where the environment is brownish in colour (moors with old bog grasses still in winter colours). When they sit on the nest with their relatively light backs, they are easier to spot than when they have a darker and browner coloration. Actively changing plumage color, as I understand it, assumes that cranes have some concept of what they look like in the eyes of outside observers.

Even when a male crane, as described at the beginning of this book, presents itself to the female after copulation in full splendour like a "federal eagle"

---

[18] Bert Brueshaber in "naturschutz heute" 4/2000 (article in German).
[19] Bert Brueshaber in "naturschutz heute" 4/2000.

symbol, it seems to have a picture of what it looks like from the female's point of view. The same is true when a crane suddenly chooses different flight paths in its territory so as to be less visible to observers. I have also described this in an earlier chapter.

If we consider this, birds could perhaps be quite hopeful candidates for the mirror test after all. In this context, I would like to talk about experiments that were carried out at the University of Bochum. Magpies, which, as corvids, are known to belong to the most intelligent of bird species, were confronted with a mirror by the researchers. Near the mirror, the scientists placed beautifully shiny and glittering objects, but in such a way that they were not visible in the mirror. The magpies purposefully brought these objects to the mirror[20] and looked at themselves there with the glittery things in their beaks—a behavior that behavioral scientists call "purposeful self-presentation with objects" and that had previously been observed exclusively in chimpanzees. The research group even observed that their magpies discovered a yellow spot that had secretly been marked on their throats in the mirror and tried to remove it.[21]

In Japan, I was told a story about cranes coming into contact with a kind of mirror. In the spring of 1987, there was a crane pair that had its territory near the property of a farmer named Mr. Otami. This pair stayed away from people, but liked to walk around near the houses in search of food. It was always seen leading a young one. The farmer's son drove a pitch-black car, which he loved to clean to a mirror finish. One day, the crane family passed near this car with the curious youngster. The cheeky youth took a closer look at the car—and spotted a crane, his reflection, in the shiny black bodywork. The young crane walked alongside the car, back and forth, and kept turning his head to look at his reflection. After this experience, the young crane came almost every day to look at his reflection (or "the crane in the car")—and one day discovered the outside mirror. From then on, he purposefully looked in that very mirror as often as he could, never pecking at it, not doing anything noticeable except looking in it with interest and turning his head back and forth. What did he see there, what did he feel?

Maybe one should do the mirror test with cranes after all.

For a long time now, I have been intensively involved with questions of thought and consciousness. That means I read pretty much everything I can get my hands on about brain and behavioral research. Time and again, I also try to

---

[20] "ND Journalismus" online 3. 7. 1999, cf. http://bit.ly/32zIN4A, (text in German, last visited Sept 24, 2021).

[21] H. Prior, A. Schwarz, O. Güntürkün et al., Self recognition in Magpies, PLOS Biology August 2008, Vol 6, Issue 8, e202 (9 pages), free online access here https://www.bio.psy.ruhr-uni-bochum.de/papers/Prior_2008.pdf

watch myself to get a sense of how I think. I pay attention to how "it" (my brain) thinks "for me" and "with me", respectively. At some point, I noticed that I rarely initially think in terms of words, but mostly in terms of images; only later, after imagining, do I also begin to think in the form of words (and sure, this mostly applies when I want to write something; not necessarily when I want to say something!). I suppose it's the same for most people, it's just that we verbalize our thoughts so quickly that we think we think in words.

August Kekulé is said to have seen his intuition about the structure of benzene in a dream as an image of snakes intertwined into a hexagon; I wouldn't be surprised if it happened exactly like that. I always formulate my theories, inventions, and solutions to problems in connection with my scientific and technological work in pictorial form first. That is, I put myself in the place of the things the dynamics of which I want to understand, and try to trace the processes from within a mini-submarine, so to speak, as an observer seemingly on the same level with the active molecules and nanoparticles. In this way, I seek to improve my understanding of the complicated processes.

I think it is quite possible that creatures that use optical impressions, such as cranes, also think in pictures, and can thus imagine future events (for example, the growth of the young, the fall or rise of the water level, the growth of the grass, or even the flight path to a destination) as "moving pictures" in front of their "inner eye". This would mean that they can visually act out problems and their solutions, as well as trajectories of their flight. The same applies to winter and spring migration. Cranes could thereby imagine a path from the beginning to the end "without words", just as we humans do (do you remember my example of your visit to Tokyo?).

Language and writing are processed in different brain regions than seeing and recognizing. The "consciousness" has nothing to do with any of these centers. As is known from clinical cases, every human being has two different "consciousnesses", one in the left and one in the right brain hemisphere. Severely ill epileptics sometimes had the connection between the two hemispheres of the brain surgically severed in the 1950s and '60s. The epileptic seizures stopped afterwards. However, these people subsequently found themselves in the unpleasant situation of being two different personalities in the same body, each uncomfortably in control of a different half of that body, and arguing with each other.

What we call "consciousness" is part of the human belief that we are superior to other living beings. But how many of our actions actually take place "consciously"? And how "consciously" do the two halves of our brain actually work together on the tasks that we accomplish day after day? How conscious is my presentation at a conference when I see images in my head, observe the reactions of my audience, and spontaneously, without interrupting myself while speaking, begin to rephrase a passage because I suddenly find it not helpful for this audience?

Every one of us has had the experience that good ideas don't come to you exactly when you need them and are specifically seeking them, but rather in relaxed situations when you are thinking about something completely different or nothing at all. I have my best ideas on my bike or in Duvenstedt Brook, or often in the first week of my annual vacation away from home.

Another example of unconscious performance is the fact that I can have an excellent conversation with my passenger while driving my car. Only in a sudden dangerous situation do I become aware of the driving itself again. Otherwise, decisions about certain movements, such as steering, accelerating, braking, are mostly made unconsciously, i.e., without "me".

High-performance athletes often train mentally, for example, simulating the high jump or the step sequence of jumping and landing in the 400-m hurdles in their imagination. I have read about pianists who also play the piano mentally. Why do they do this? Because then, when it comes down to it, they can let their activities run, to a large extent, unconsciously, quasi "automatically".

So, there are aspects of life in which we get along quite well without consciousness, and are even more efficient without conscious control, or perform the tasks set better than we do with it. This applies, of course, to all functions of the body that are controlled by the autonomic nervous system, which functions "unconsciously" anyway. We cannot consciously control heartbeat or gastrointestinal activity (and can only partially observe them) and would be quite overwhelmed by all of the activities and stresses of everyday life if we also had to think about pumping our blood and digesting our food. It is similar with breathing, which normally takes place unconsciously; we can inhale and exhale consciously, but bad breathing habits often creep in (unconsciously) due to stress, which, in turn, can be remedied through (conscious) breathing training.

What I find interesting is the phenomenon that many people (including myself) can make themselves wake up at an unusual time in the morning, for example, if they have an important early appointment. Without a doubt, when we sleep, we are without consciousness, and waking up happens without our conscious control. One does not even have to have made it a point to wake up on one's own, perhaps having set the alarm clock, but the (unconscious) setting for what may be a crucial early morning matter turns out to be enough. On the other hand, there are many cases in which it helps us to "turn on" our consciousness. My point is not to attribute exclusively a disruptive influence to consciousness—quite the opposite. I believe that we should use our consciousness much more often, especially in observing and critically assessing ourselves.

Everyone has certainly noticed with themselves or others instances in which they have said something in a discussion that they would rather not have said or even said something completely different that "just slipped out". The tongue was faster than the brain, or, at least, than the conscious and self-critical part of the brain.

Certainly interesting, and arguably confusing, is recent research that suggests that the decision to do something seems to occur in the brain sooner than we become aware of it.[22] I was not surprised by these results, because I have often felt that my brain thinks with me (or for me), that is, that "my brain" and "I" may well be two different subjects. If one observes oneself and other people, one may find that "consciousness" varies subtly from person to person, as well as within a person over time. All in all, "consciousness" is too complex a phenomenon to have been deciphered by psychologists, biologists or biochemists so far.[23] I am convinced that consciousness has developed evolutionarily and was not simply "there" when the first Homo sapiens, i.e., the first "rational" human being, appeared.[24] But this also means that there must have been different stages of consciousness development in the course of the evolution of all animals. So, I would not be surprised if different creatures—for example, monkeys, wolves, elephants, salmon, ants or cranes—do not differ simply in whether they have consciousness of themselves or not, but rather in what degrees they have it and what kind of consciousness they exhibit.

I don't have a precise plan for finding out more about this, but one method would certainly be to observe living creatures in their natural environments to

---

[22] https://www.spektrum.de/alias/r-hauptkategorie/hirngespinst-willensfreiheit/968930, compare also https://www.spektrum.de/news/wie-frei-ist-der-mensch/1361221 (last visited on Sept 24, 2021), all texts in German, cf. English overview in https://en.wikipedia.org/wiki/Neuroscience_of_free_will#Libet_experiment and publications by Benjamin Libet cited therein.

[23] For a first introduction to this complex topic, this article is quite suitable (however, it is in German): https://www.zeit.de/zeit-wissen/2012/02/Mensch-Individuum-Selbstbewusstsein/komplettansicht; the scientists cited therein had been the research group in the "ChildLab" of the Max-Planck Institute for Evolutionary Anthropology (https://www.eva.mpg.de/comparative-cultural-psychology/research-infrastructure/ccp-child-lab/), Michael Tomasello (https://en.wikipedia.org/wiki/Michael_Tomasello), Michael Gazzangia (https://psych.ucsb.edu/search/node?keys=Michael+Gazzaniga), Olaf Blanke (https://www.epfl.ch/labs/lnco/olaf.blanke/), and others.

[24] In 1976, Julian Jaynes published a book with an interesting thesis: What we today regard as "consciousness of ourselves" only emerged about 3000 years ago. He derived this from the analysis of historical texts. His book (I read the German edition: Der Ursprung des Bewusstseins, Reinbek: Rowohlt 1993; original English edition: "The Origin of Consciousness in the Breakdown of the Bicameral Mind", Houghton Mifflin 1976, most recent edition Mariner Books 2000) is well worth reading and makes clear how "consciousness" could have developed, and affirms that "consciousness of oneself" is also a result of evolution, and, as far as humans are concerned, possibly even one that only began to exist within recorded history.

see how they deal with unknown problems. I've talked a lot about that over the course of this book. But amazing observations have been made not only with cranes, but also with other birds. For example, in New Caledonia, there lives a crow species that makes and uses tools. The birds bend twigs and work them to make fine stems with hooks, which they use to fish for prey in burrows and holes. The first tools used by humans were more primitive.

I think all problem-solving that requires foresight is only possible with some degree of awareness. Only in this way can the living being in question imagine a future situation (pictorially, for example) and play through alternatives.

I am particularly fascinated by the ability of animals to actively hide. I mentioned the great spotted woodpecker hiding behind the tree. I told you about the crane flying in front of or behind a row of bushes, and about how a crane faked a duet call to me.

The ability to do this seems to me to be related to what is called the "theory of mind". This is the knowledge that other beings have a view of the world that differs from one's own—an ability that is generally only attributed to humans, or, at most, to chimpanzees. Indian biologists have discovered evidence of this ability in a tropical species of bee-eater.[25] In observing these birds, they found that they were reluctant to reveal the location of their nests, so they always tried to avoid flying there when a researcher was in sight. In one variation of the experiment, the biologist's view of the nest (but not of the bird) was obscured. Amazingly, the birds then flew to their nests much more often: They could see the human, but could also imagine that the human was not able to see the nest from their vantage point. Being able to see the world from another's point of view goes far beyond what behavioral scientists had previously thought birds could do.

In this context, the so-called "seduction" also comes to my mind: In this process, ground-nesting birds (such as lapwings or golden plovers) distract a potential nest or chick predator by flying towards it and playing the injured bird by limping and seemingly desperately trying to escape with their wing hanging down. In this way, the intruder's attention is diverted and the bird pretty much guides it away from the nest or chicks. Is this really an innate behaviour or do the adult birds know what they are doing, having picked it up from their parents in their earliest chick days?

---

[25] B. Smitha, J. Thakar, M. Watve, "Do bee-eaters have theory of mind?", Current Science 76 (4): 574–577 (1999), full paper available here: https://www.researchgate.net/publication/289872605_Do_bee_eaters_have_theory_of_mind

Again and again in these reflections, I come back to the question of what the brain actually is, and so far, I have not found a satisfactory answer. In any case, it is interesting that the brains of all higher organisms are much more similar to each other than, for example, their digestive or sexual organs.

One thing seems clear to me, though: Brains are unique. And as much as the development of "artificial intelligence" is progressing, a closer look shows that a brain does much more than what we understand when we use the word "intelligence", and manages much more complex tasks than any supercomputer.

Roger Penrose is one of the leading physicists of our time, and he constantly argues against the viewpoint and the wording according to which computers and certain special software are said to be "artificial intelligence". Of course, he does not doubt that computers can do certain things much faster than we can with our brains. These are questions for which the path to the solution can be mathematically formulated and standardized in so-called "algorithms".

Penrose has shown that humans, on the other hand, can also answer questions that are clearly not computable with the help of algorithms. He refers, for example, to certain mathematical theorems that cannot be proven by algorithms, but are nevertheless correct. Penrose calls these questions "non-computable problems". A computer would, in principle, be overwhelmed by them. Essentially, this has to do with "understanding (Begreifen)".[26]

From the fact that the human brain can also solve "non-computable problems", Penrose concludes that it must be structured differently and function differently than we currently think. It cannot simply consist of an extremely large—in itself already impressive—number of connections between brain cells, but must be based on processes that are also not computable as such. In addition, we can be fairly certain that the brain does not operate in the same manner as computers, with links that are switched between "zero" or "one",

---

[26] Roger Penrose, "Im Schatten des Geistes", Spektrum Akademischer Verlag 1995, et al. p. 82 (German Edition) (original English edition "Shadows of the Mind: A Search for the Missing Science of Consciousness", Vintage Publishers 1995); consistent with this idea is the fact that apparently clever "deep-learning" programmes that beat the best in the world at the game of Go have no understanding of numbers whatsoever and cannot, for example, calculate "1 + 1 + 1 + 1 + 1 + 1 + 1" correctly, but come up with "6" by mistake: https://www.spiegel.de/wissenschaft/kuenstliche-intelligenz-in-der-krise-zu-dumm-zum-rechnen-a-00000000-0002-0001-0000-000166735218; so far, no AI has been able to "understand" what can actually be seen in a recognisable photo, sometimes making glaring mistakes; no AI "understands" a human conversation. As long as everything runs normally and the rules don't change (like in the board game Go), AI can be used, but even the smallest surprises lead to absurd mistakes. A computer can do useful, routine things very well, but in the real sense of understanding, it doesn't really know anything about what it is doing.

"yes" or "no". Rather, states in the brain seem to build up in three dimensions, flowing through each other, and thereby allowing a great many processes to occur in parallel and intertwined. Exactly how this happens must remain a matter of speculation, even for Penrose; however, he has conclusively proved that a brain is qualitatively more and can do more than any computer.

From my point of view, this is already true even for a "sparrow brain". Once, I stepped around a path bend and stopped because I wanted to observe a group of sparrows sitting on the path. One of the birds saw me and gave a warning call, whereupon the entire troop flew into the bushes—except for the warner, which remained sitting calmly, grabbed the biggest crumb, and hopped into the adjacent ditch. A successful deception, organized by a brain weighing only a few grams—with a flexibility that beats any artificial intelligence by far.

Granted: It is a long way back from my thoughts about the consciousness and intelligence of humans to the cranes. But in the following, I would like to try to bring the observations I could make on cranes in line with my thoughts about "consciousness".

Since we humans, as has been pointed out, do not always act "consciously" either, we must not assume that "consciousness" is something that can be switched on and off, so to speak, something that is either completely present or not present at all.

In this respect, even the highest degree of consciousness that we humans grant ourselves cannot actually be rated as "100% consciousness". The truth lies somewhere between 0 and 100—let us take, as an example, a human who is awake and perhaps driving a car and form an estimate: this person is 65% "conscious at it". His degree of consciousness may be higher in certain situations, perhaps when challenged with a complicated task in his professional work, and will certainly be lower in other situations, perhaps when taking a rest in his backyard in a sun lounger under a tree, doing nothing more than watching the leaves and the clouds; not to mention a zero degree of consciousness while sleeping at night.

Once we have taken this mental step, it will be easier for us to accept that the "consciousness of oneself" may also be gradually different in different animal species (and thus in different individuals of a species). Presumably, we agree that no species of animal achieves "65% consciousness". But perhaps one or another species may achieve 40%, still others twenty?

Biologists and behavioral scientists working on this topic have created a categorization according to which there are two levels of "self": one called the "machine-like" (or "mechanical") self, and another linked to a "conception of one's own self". Scientists assume that only those animal species that recognize themselves in the mirror possess the latter.

I think this restriction is pretty anthropocentric. Moreover, it equates awareness of oneself with the ability to recognize oneself in a mirror (thus requiring only visual perception and, in addition, that the inner image of oneself matches that in the mirror). This excludes the possibility of recognizing oneself acoustically or by smell.

Some scientists go even further, demanding "language" as a prerequisite for the existence of a conception of the "I myself". Because only then can this being speak of "Myself" and communicate whether it has an idea of its own and how it differs from others. With this definition, all living beings without the ability to speak are equally disqualified—a dog as well as a slug.

Undeniably, human language enables a complex form of communication. But what do we know, for example, about the language of whales and dolphins, or parrots? What do we know about the non-verbal communication capabilities of animals when we ourselves do not even know exactly how much of our communication consists of non-verbal signs within our body language? Many studies show that we are overwhelmingly unaware of the body language we ourselves use. Psychologists even believe that the greater part of communication among us humans is not verbal, but is represented through facial expressions and gestures, largely unconsciously (!). What's more, just recently, US scientists compiled thousands of non-verbal vocalizations and the total of 24 different emotional messages associated with them into an interactive map.[27] Now, it has been scientifically backed up: We can communicate emotions without language. Why, then, could not animals do the same? Why not birds? This is an important question, because the capacity for so-called "primary emotions" is described as a precursor to the development of consciousness.

It is not the aim of my considerations to prove that cranes are as gifted with consciousness as humans and to classify them as philosophers pondering over their lives. But perhaps there is a certain degree of (self-)consciousness in cranes or other birds, and also in mammals—even if it were "only" 10% or 20%.

The evolutionary biologist Marc Bekoff and the bioethicist Jessica Pierce have written a book with the title (German edition) *Sind Tiere die besseren*

---

[27] cf. https://news.berkeley.edu/2019/02/04/audio-map-of-exclamations/; the original publication can be read here: https://www.researchgate.net/publication/329824563_Mapping_24_Emotions_Conveyed_by_Brief_Human_Vocalization. To my astonishment, our non-verbal expressions of affirmation and denial or agreement and disagreement, which animals may also have at their disposal for communication, are missing from this map, and I am even more dismayed by the absence of sexual "lust" and "desire", which we humans probably also communicate almost exclusively non-verbally.

Menschen? ("Are animals the better humans?")[28] They argue that animals are more like humans than we think. The neuroscientist Antonio Damasio is of the opinion that it was not the mind and intellect, but rather emotions that were decisive for humans to become human beings.[29] He wrote that, while our limbs, our physiques and brains, our organs, and our senses clearly developed over the course of evolution, there is no reason to assume that feelings, emotions and consciousness appeared suddenly and only in humans. My many years of observing cranes in so many parts of the world have led me to believe that cranes have a diverse emotional life. This book's attachment contains a comparison of a range of human emotions with my various observations of cranes. However, many scientists believe that most animals do not have a consciousness because they lack a memory that stores events in order of space and time and keeps them available for experiences and decisions (this is called "episodic memory"). I will address this question in the next chapter.

I conclude this chapter with an assertion that I would like to see as a suggestion for further observation and research: Cranes possess a high degree of intelligence. They are able to solve problems that occur flexibly, i.e., adapted to the situation. They have complex communication skills and a certain awareness of themselves. The question as to what extent they can develop a differentiated conception of themselves must remain open. I would not put the awareness of themselves at 0%, nor at 40% or 50%. It will be somewhere in between, perhaps comparable to that of dogs, wolves or bears, but based on a different brain structure, namely, the bird brain.

The idea of the "I" (or self) develops by benefit of a social life with conspecifics. In cranes, as socially living animals, a conception must have developed that the other one is a someone, but just another someone. I suspect that the self-image and the image of the "others" in cranes is pictorial and probably "tonal" at the same time.

---

[28] Marc Bekoff, Jessica Pierce: Sind Tiere die besseren Menschen? Stuttgart: Kosmos, new edition 2017; original English book: Wild Justice—The Moral Lives of Animals, University of Chicago Press 2009.

[29] Antonio Damasio: Im Anfang war das Gefühl—Der biologische Ursprung menschlicher Kultur. Munich: Siedler 2017; English edition: "The strange order of things: Life, Feeling and the Making of Culture", Pantheon Books, 2018.

Can Cranes Think Strategically? More Amazing Observations

Consciousness in cranes—not everyone will be congenial towards the thoughts laid out in the previous chapter. I can understand that very well, because it took me several years before I conceded that cranes have at least some percent of consciousness of themselves, and with increasing observations, that percentage has risen higher and higher.

I would now like to turn to another topic that I touched on briefly towards the end of the previous chapter: the so-called "episodic memory". This refers to the ability to remember experiences that oneself has had at a certain time within a certain context. This therefore requires that the being that has an episodic memory "knows" that it itself was present at the event underlying the remembered experience, that it itself had this experience at that time at that certain location (or observed someone who had an instructive experience—but that would be an even higher level of ability for episodic memory).

The vast majority of behavioral scientists believe that birds do not have episodic memory. This view has long been challenged, if not disproved, by research on the American scrub jay *Aphelocoma coerulescens*. These birds like to hide their favorite foods, maggots and peanuts in the case of a particular controlled experiment. The maggots, however, only taste good when fresh, while the peanuts last longer in the hiding places. How well the episodic memory of these jays works was proved by the fact that they knew exactly how long it was still worthwhile to visit a maggot hiding place. If they did not have the chance to seek out fresh maggot hiding places, they left them and ate the peanuts. But if they had a chance to dig up the maggots soon after hiding

© The Author(s), under exclusive license to Springer Nature Switzerland AG 2022
B. Wessling, *The Call of the Cranes*, https://doi.org/10.1007/978-3-030-98283-6_14

them, they preferred them.[1] There is no doubt that the scrub jays can selectively seek out more than 1000 different hiding places over a long period of time, relying solely on their memory of their location and age. However, this still does not come close to what I could observe in cranes, namely, strategic planning. This requires, in addition to memory (what happened when, where, in what context), the ability to recognize that a current problem is similar to the situation at that time, and the ability to draw the conclusion of what to do instead of the action (or omission of an action) at that time.

In the course of my observations over many years, I found abundant proof that cranes have such abilities, i.e., they can process experiences made earlier and mentally prepare a solution for problems that occur later. At least, I do not know of any other explanation for some of my observations. But again, I am fully aware that these observations are not reproducible; they have not been made under controlled experimental conditions, but in the field.

In the spring of 1996, for example, the crane pair that had been breeding in a particularly wet part of the "Old Moor" moved back into their territory and felt completely at home there, as before. They searched for food, mated and prepared for breeding—sometimes on the meadows next to the nesting area, sometimes far away on the edge of the brook.

Three years earlier, the two had conquered this territory, which closely bordered that of a pair already established there. The "negotiations" about the exact borders between the territories lasted for weeks. For this purpose, both males walked the boundaries very demonstratively, with their heads raised high and their necks stretched ("brag march"). Among other things, they agreed on a ditch and a sparse row of bushes as a boundary (thus, presumably, using visual markers to be able to remember their own and the other's territory). These boundaries were then accepted by both sides and were no longer contested, and we never again observed territorial disputes between the two pairs.

However, instead of enjoying this peace and quiet in the spring of 1997 and leisurely preparing the brood as in the previous year, the cranes behaved differently, and strangely, this time. Whenever the pair was relatively close (up to about 200 m) to the future feeding meadow for the young, the male scared away all the geese that were there. Over the course of the first few weeks, it became apparent that he only tolerated them beyond an older ditch, a line visible to cranes and geese. This was strange, because, in the 3 years that this pair had been breeding there, they had always lived in peace with the geese that were also breeding there.

---

[1] cf. https://www.nature.com/articles/26216, In this article and in general, this ability is only referred to as "episodic-like memory", i.e., is only considered "similar to human episodic memory": https://en.wikipedia.org/wiki/Episodic-like_memory. Why this restriction?

In the past, the geese and the crane family had been seen close together in the grass—one eating, the other resting or preening their feathers. What was different now? Why did the crane no longer want to tolerate the geese? It can't be because of any competition for the same kind of food.

The hatching went off without a hitch. The water level was relatively low from the beginning, with the little rain that fell hardly changing it, and it remained quite dry. The cranes had chosen a favourable spot for their nest, so it was and remained safe.

After hatching, the crane parents appeared in the meadow for the first time one day with their two tiny chicks. A couple of geese ventured over again. They should rather not have done so; they were immediately chased away. At that, apparently, the previously valid boundary was no longer sufficient. The crane drove the poor geese away to the next meadow but one. Strangely enough, it left the still very small chicks alone with their mother (normally, when two have hatched, one parent takes care of each young one until they have become older). So, it must have been very important for the cranes to keep the geese at bay. But why?

For weeks, it went on like this. The radius around the cranes, in which the geese were not allowed to set foot, became increasingly larger. Soon, they were no longer allowed to venture closer than 300 meters. It had long been clear that the geese were having a disastrous breeding season: Not a single brood was successful in the area; we didn't see a single gosling. The young cranes, on the other hand, were growing rapidly and would soon fledge. The reason for the constant goose evictions, however, remained unclear to me.

One day, I became aware of the crucial question: Had I actually seen the fox that usually quite frequently hunts there, if only just once that year? No. I asked all the fellow crane-keepers and observers I could reach. No, no one had observed the fox. Was that the background of the crane's strange behaviour? The fox had certainly not come by there in the past for the maximum of one or two crane chicks potentially to be captured, but instead for the 35 or so geese pairs and their 20–40 goslings. That had often made the hunt worthwhile. But, incidentally, he had certainly bagged one or two crane chicks from this pair every season for the last 3 years; once, we had even observed it.

Had the cranes noticed this? Had they noticed this connection? Had they perhaps realized that the fox was actually hunting geese? We don't know, but the effect was, as described, that the fox hardly showed up in the crane territory, or even not at all, because of the lack of geese and goslings. And two splendid crane chicks had fledged that summer. That was unique. The incident is, I think, a strong indication of episodic memory and the ability to plan strategically. I named this crane pair (M4F4) the "Foxes" as a result of this incident, because they were cunning like a fox.

In the following year, the crane did not hunt the geese as intensively, but still to a sufficient extent. Again, the geese could not breed at this place (at least, we did not observe any gosling). Although, every now and then, a few geese would graze and spend the night near the crane nesting location or territorial boundaries, they moved to other areas in the Brook to breed. Even today, the geese no longer breed in just one place, but in several different places. In the meantime, they have returned to the meadows mentioned above, as new generations of cranes and geese have been living there for a long time.

Another convincing example was provided one morning by the Texas Whooping Crane pair described earlier, which, to my surprise, I once found not at their roost in the water, but near an area where fire had smoldered the day before. They had spent the night on dry ground, contrary to crane habits and all the usual crane safety regulations so they could pick up the freshly grilled goodies in the morning. There was obviously some clear planning behind this. Consider this: The observation "There's smoke" was linked to "Where there's smoke, there's fire," and, further, "Where there's fire, there will be grilled goodies to be found a little later". This is already a decent amount of quality cognitive achievement! And then, to a certain extent, as icing on the cake, the conclusion: "We have to spend the night there today and not on the coast in the water, so that we'll be there first in case other cranes come by."

Each observation of cranes that visit certain fields once and then repeatedly if something nutritious is to be found there testifies to episodic memory, as well as the shifting of brood, resting, gathering and wintering places.

Strategic planning is closely linked to the ability to decide to do something in the face of an alternative. This is called "free will". Marc Bekoff and Jessica Pierce also discuss this aspect:[2] First, they make clear that animals can and do have free will, and that, in a way that is astonishing to many of us, voluntary decisions in the animal kingdom are often made from a moral point of view ("right"/"wrong" or "acceptable"/"not acceptable"). On the other hand, moral behavior may well be conditioned or instinct-guided, which is also often the case with humans. In my opinion, the concept of "free will" does not necessarily involve moral components; for example, the decision not to fly to this meadow this morning, but rather to fly to that field, is "amoral", that is, morally neutral. So, cranes do indeed seem to have free will. Last but not least, we can see this in the fact that they can decide on different migration routes (and

---

[2] Bekoff/Pierce, op. Cit. (German edition) p. 205–207.

actually change them), and even spontaneously decide whether they want to spend the winter in northern Germany, as long as it does not freeze too violently and for too long.

But do we also find indications in cranes that they have at least the beginnings of a moral code? Bekoff/Pierce have devoted part of their research to the question of "moral codes" in the animal kingdom.[3] They found that, in social animals, there are very clear codes by which "wrong" and "right" are judged, i.e., an action is not accepted/punished or is accepted/rewarded. They write that this "might even be the case in birds". Can we see this in cranes? Yes, I think so, for example, in territorial behavior: Once it is clear who the territorial owner is and what their territory encompasses, the unison call or, if necessary, the much more elaborate minute-long "brag march" (still an unexciting act) is enough to reinforce that clarity. Pairs that have already had their territory next to each other for more than a year will no longer need territorial disputes the following spring; it is clear to both sides that crossing the border to the other side is "wrong", "that is not right!"; one respects the other's territory. New couples, especially young ones, have to learn this. I would certainly consider this a moral code.

Normally, territorial fights are bloodless. By "normal", I mean "in more than 99.999... percent of the cases". Only in very rare cases, which are then reported about excitedly and often with photographic evidence, does a territorial fight end bloodily (I have never observed this myself, although I have seen hundreds of territorial disputes). From this, I conclude the following: On the one hand, cranes have occasionally, very rarely been observed to fight each other to the death, but on the other hand, they hardly ever do so. Here, I think, we see the result of a kind of free will, a free choice to acknowledge a moral code that might go something like this: "Territorial fights, yes, but we only attack in pretence; we demonstrate strength, but only extremely rarely do we actually use it!"

Admittedly: I inject a lot of interpretation into my observations. The observations are facts, the interpretation I like to leave to the individual. I myself go as far as I do because I can think of no other, equally logical interpretation.

I do not mean to suggest that cranes are more intelligent than corvids or parrots or pigeons. The purpose of this book is not to trump the unique work of one Irene Pepperberg with her parrot Alex. That one could talk, understood

---

[3] Bekoff/Pierce, op. Cit. (German edition), there is an easy-to-read review article in English by J. Pierce, M. Bekoff at: https://www.chronicle.com/article/Moral-in-ToothClaw/48800/, last visited on Sept 24, 2021.

the questions of his human caretakers, and was able to answer "reasonably." Her research results are sensational.

I don't know whether "our" cranes, like the laboratory pigeons of some behavioural scientists, can tell colourful triangles, squares or abstractly painted trees and houses apart and sort them out in order to get their food when they assign them correctly. They've never practiced it, and I'm certainly not going to train them in it, because, in the Brook, in Siberia, in Sweden, in northern Canada, or on migration to the south, they don't have to sort abstract symbols to find food. They have to be able to distinguish snails and worms from branches and roots, corn kernels from colorful pebbles, dry areas where nothing grows and they can't spend the night from worthwhile fields and meadows to feed in. They must be able to recognise wetlands suitable for breeding and shallow waters for the night. And they must be able to distinguish dangerous foxes from badgers, as the latter do not hunt young birds but are very fond of slurping up eggs, so they should not be permitted to get at the nest.

I don't know if cranes exchange information about food and environment like corvids,[4] but I assume so (please remember my previous mention of the emergence of new wintering or resting locations!) and they can definitely summon each other. I doubt that they would secure thousands of hiding places where they have hoarded acorns for the winter like jays do. Cranes don't gather acorns or corn kernels and hide them for the winter. They look for food when they are hungry and where it can be found, even if that means flying to Spain or Ethiopia. Conversely, jays would never make their way to Spain, nor would I myself without a map and road signs.

We humans have defined quite narrow criteria for "intelligence": If animals can do the same thing that we can, we consider them "intelligent" in that respect. If they can do something that we cannot, it is considered "innate". Perhaps it would be better if we cautiously considered all behaviors that are not verifiably innate as "learned" to begin with (at least as a viable hypothesis), because, all too often, an allegedly innate behavior has turned out to be surprisingly flexible and arguably learned after all. Behaviors that enable animals to survive, to obtain food on a daily basis in a wide variety of situations, and to survive difficult weather conditions are highly intelligent. Such flexible

---

[4] Cf. https://www.zeit.de/2007/26/N-Raben/komplettansicht, an interesting article in German—readable after registration—that has the same topic as chapters "What Can We Learn About Intelligence, Migratory Behavior, the Formation of Culture, Tool Use, and Self-Awareness in Cranes?" and "Can Cranes Think Strategically? More Amazing Observations". The focus is not on wild cranes, but on common ravens in a research institute. Cf. also https://www.welt.de/kmpkt/article178370042/Raben-reden-gezielt-miteinander-und-rufen-auch-um-Hilfe.html (articles in German, last visited on Sept 24, 2021) (an article in English linked to this topic can be found here: https://frontiersinzoology.biomedcentral.com/articles/10.1186/s12983-018-0255-z#Bib1).

behaviors are species- and individual-specific because different species and even individuals of the same species live in different ecological niches and have different evolutionary histories behind them, thus they have different physical and mental prerequisites. In addition, they have accumulated different (partly individual, partly cultural) experiences.

Cranes have lived through and helped to shape 60 million years of the earth's history, have found their way in a wide variety of landscapes and climates, and have successfully dealt with animal species that, for the most part, no longer exist today. The behaviours that we observe today and the brain structures that have enabled them to engage in such behaviours have emerged in the course of their evolution—on the one hand, the physical and mental foundations, on the other, cultural traditions and personal experiences or learning outcomes or unique brain performances ("decisions", "inventions").

We should therefore regard as intelligent behaviour all that which is performed on the basis of a brain performance as a flexible response to changing circumstances and challenges in order to survive and reproduce. This is much more than is captured in intelligence tests and laboratory experiments.

The examples of planning or long-term strategy described above and in the previous chapters are particularly impressive, and cranes like the "Foxes" are not unique. But cranes are obviously even capable of tactical behavior that would do credit to a soap opera script.

In an easily observable territory on Hokkaido, there lived, for many years, a crane pair that had successfully bred and, as could be seen from their rings, was already quite old. At the end of March 1995, the banded female "F15", only one and a half years old, suddenly appeared. She drove away the old female of the territorial pair and made an alliance with the male landowner. He agreed, and the two spent a few exciting weeks together, of course, without laying any eggs, because the young female was not yet sexually mature (the old male probably did not realize that, or didn't care).

In June, the crane conservationists observed how "F15" suddenly drove the old male away. Of course, he kept coming back, wanting to reclaim his territory, and certainly wanting to keep the attractive young female, but she mercilessly scared him away, apparently wanting to be alone. Finally, he gave up. A short time later, the young female was seen in the territory feeding with a new male crane, also still young (also banded), side by side in harmony. The observation logs on the feeding areas in the previous winter months showed that "F15" had already been with a crane of the same age 6 months earlier.

It was not until 1998 that the two had their first brood. The conquest of the territory obviously took place after some very skillful, long-term planning; if humans had behaved similarly, we would call it "a perfidious strategy". If the two inexperienced young cranes had engaged in a territorial fight with the experienced pair, their chances of taking over its territory would probably have been much worse.

Episodic memory, planning ability and, in addition, a further, even more highly developed cognitive ability are revealed by the following observation by Ukrainian crane researchers: Sergej Winter and his colleagues observed a total of 70 crane territories over several years in two areas and investigated how many nests the breeding pairs built. They found that about 40% of the pairs built two or more nests in their territory, and that about a quarter of the cranes built nests but did not lay a clutch and did not reproduce. It was these pairs that built two or more nests, up to five. I had also noticed this phenomenon on several of my nest searches in the Brook. However, I had not realized that these were apparently not exceptions or the byproduct of a quirk, but a habit with an unknown background. Perhaps the construction of multiple nests is for testing nest sites or to mislead potential nest predators. Interest in a game ("We're playing brood!") or boredom could also play a role. Sergej Winter and his colleagues rather suspect an instinct-driven background and speak of an "exaggerated motivation for nest building".

I met Sergej Winter when I presented my then-first observations on crane behaviour at the 1996 International Crane Conference. He approached me after my talk and told me something that he had not mentioned in his own talk, which had preceded mine:

He and his colleagues had noticed one crane pair in particular. It was a pair that bred in a wide wetland, which was predominantly covered with reeds and sedges, as well as black alder and low willows. There was no meadow or clearing anywhere in the vicinity of the nest for the chick to roam in. The pair had then built a "playground" for it: They had folded down or torn out the tough reeds, which would have formed an impenetrable thicket for the young cranes, and laid the stalks flat on the water, thus creating a flat, free surface on which the chick could run around without any problems (Graphs 1 and 2).

After Winter and his colleagues discovered this, they found such playgrounds everywhere that similar conditions existed. Depending on the conditions at the respective crane pairs' breeding sites, these playgrounds looked very different. One of the pairs, at whose nest the stalks of bank sedge were not very dense (four to eight stalks per square metre of water), laid all the stalks flat on the water, covering a total area of 20 by 40 m. Another built a playground of 20 by 30 m on a section with a high density of vegetation (30–50 culms/m^2). Incidentally, the earliest time to build the playground was 2 days before hatching, when the chick in the egg was already peeping.

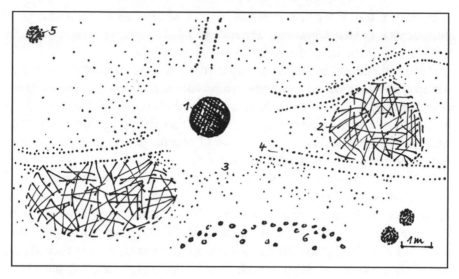

Graph 1 Playgrounds established by Eurasian Cranes in Ukrania: Scheme of nest environment and locations of the "playgrounds" of the grey crane (sketches according to Sergej Winter): (1) nest, (2) "playground" with grass laid on the water, (3) grass cover, (4) paths made and used by the cranes, (5) single black alder, and (6) dense willow scrub

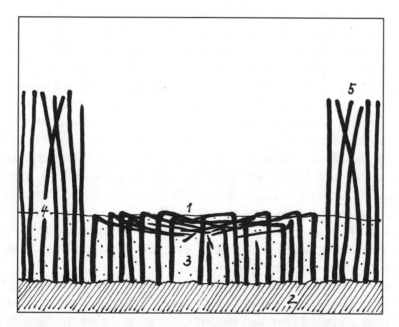

Graph 2 Playgrounds established by Eurasian Cranes in Ukrania: Scheme of nest environment and locations of the "playgrounds" of the grey crane (sketches according to Sergej Winter): (1) plants laid on the water surface, (2) muddy bottom of the flooded black alder grove, (3) water, (4) water surface, and (5) grass cover, predominantly bank sedges

In 1990, at least five crane pairs had established between one and four playgrounds each in that Ukrainian wetland. In 1991, there were eight pairs, in 1993, five again, as far as the scientists could determine.

Had a real culture developed here, passed on as a tradition of the cranes in this area, one that we do not know from other areas? I would like to know when the playgrounds were invented by Ukrainian cranes. Will we find them in Finland one day? I find it very remarkable that the playgrounds are made ready just before hatching—what a feat of planning, one that presupposes that the cranes know the time of hatching!

As the French behavioural scientist Emmanuelle Pouydebat points out,[5] the construction of "structures" (such as the dams of beavers or the individually very creatively constructed and extremely skillful nests of bowerbirds) can be equated with the intelligent use of tools. Not so long ago, this represented one of the criteria that set humans apart from the rest of the animals. In the meantime, hundreds of animal species (by far, not only primates!) have been observed to use or even manufacture tools—cranes have not been counted among them so far, because their constructions, the playgrounds, were not yet known. Here, we observe, besides the production of a special artificial facility (synonymous with tool production and use), episodic memory and strategic planning on the one hand and a cultural achievement on the other, because the creation of such playgrounds is clearly not inherited, but started with the efforts of a single first pair in an area. That was a real innovation!

Another example that should inspire us to think more deeply: Every autumn, Zsolt Végváry, an employee of the Hungarian nature conservation authorities in Hortobágy National Park, carries out counts of migrating cranes. The best time to do this is in the evening, when the cranes descend into the shallow water fishponds on the edge of the national park to roost.

Végváry and his colleagues have repeatedly observed that large crane groups approach from afar, but do not fly directly to the fishpond. Instead, they make a stopover in fields or meadows about a kilometre away from the waters.

A few cranes, however, continue to fly. They fly over the waters that are being considered as overnight roosts, circling, seeming to scout them out. Then, they return to the group. Soon after, the entire group takes off and settles in the water for the night.

In view of this observation, the following questions arise in my mind: Have the cranes flying ahead been sent as "scouts" for safety? What are they supposed to be checking? How do they communicate to the others that all is well?

---

[5] Emmanuelle Pouydebat: Was Tiere können, Munich: Goldmann 2019, p. 142 ff., original edition in French: "L'intelligence animale", Odile Jacob 2017 (to my knowledge, no English edition).

What would they do if that were not the case, that is, if fishermen were seen on the pond or anything else that was strange to them?

Already during my own observations of the cranes, it became clear to me that every new insight in turn leads to new questions, for example:

Wouldn't it be more logical to assume that, in the course of evolution—instead of rigid, instinct-driven behaviour—the development of living beings with brains capable of learning and making flexible decisions, with a certain degree of consciousness, resulted in greater success for the preservation of the species? If, in the course of evolution, both paths—behavioural control through, respectively, heritable instincts and experience and cultural learning—were developed, the latter is likely to have predominated due to its higher adaptability. That's what I would assume.

The stories that I witnessed being enacted by the cranes both amazed me and inspired me to far greater contemplation. The more we learn about cells, genes, the brain and evolution, the more mysterious and wonderful nature seems to become. In the laws of nature, in life on earth, in evolution, and in the universe, there is power and diversity that we still cannot measure.

Cranes Are Subjects. A Plea for More Modesty and Respect for Nature

The longer I worked with my recording technique, the more surprising things I observed and the more insights I gained, the more deeply I felt: What a privilege to eavesdrop on wild cranes and to learn to understand them better and better. What a gift of nature. I owe numerous moments of happiness and the opportunity to find deep relaxation while learning so much about nature, about the world we live in, to my involvement with the cranes.

I experienced a generational change in the Brook. My old acquaintances disappeared bit by bit; only the "Foxes" were still there in 2007. I had met them personally in 1998, the very first year of my recordings; presumably, they had already been in the Brook for several years. The same was true for the female "Planner", the male "Rattler" and the "Seekers". The "Pioneers", first identified on a 1996 video, were last identified in 2005 (although each with a new partner). New crane pairs arrived and had lives as eventful as those of the old familiar ones; the cycle of nature, coming, becoming, living and passing, played out before my eyes and ears in a dynamic crane society. After 10 years of getting to know all of the cranes personally, the composition of the Brook's crane society was entirely new except for one pair.

Of course, I still only understood a fraction of the life of the cranes in the Brook (and the same is true for the Whooping Cranes), but it was so much more than what we had known or understood before. Whenever I witnessed a pair answering the call from my megaphone, whenever I spontaneously recognized a crane by its call, I felt a deep happiness and gratitude. This is true even now. Ultimately, it is these feelings that have driven me to currently begin recording and analyzing crane calls again. I was and am grateful for the permission to listen and understand even this little bit.

B. Wessling, *The Call of the Cranes*, https://doi.org/10.1007/978-3-030-98283-6_15

Cranes have been around for millions of years, 60 million years, it is said. Scientists have found that cranes, like all other birds, are, in a sense, modern, more evolved dinosaurs. It's only been a few million years that there have been human-like creatures and, eventually, since less than 300,000 years modern humans. For a few thousand years, people seem to have been thinking about how cranes live. They appear in mythology, and in Asian culture, they are symbols of fidelity, eternal life, and good luck.

Over thousands of years, nature has kept its secret and has not let us humans look behind the scenes; if one does not want to assume any intention, one must at least say: The connections are so complicated and the details of the life of the cranes so mysterious that no human being had previously been able to peek under even the edges of their secrets.

That is, until people turned cranes and their lives into passive objects of their research, banding or radio-tagging them. You can spin it however you like: From that moment on, humanity was no longer an astonished, humble and modest observer of nature, watching and listening, but rather an intruder, massively intervening and robbing the crane subjects of their independence. As you know, I am not fundamentally opposed to banding and tagging. It is important to obtain additional data for a deeper understanding. But this does not change the fact that both procedures are based on a different philosophy than I practice: Banding and tagging turn subjects into objects, and they reinforce our false human assumption that we are above nature, that we can manipulate and tame nature according to our will and for our benefit.

I leave the cranes their independence, resolutely choosing not to disturb their plans and activities; I want to get to know and respect them as subjects. And so, they only reveal as much of their lives to me as they wish to. They don't have to call, but I know: When they do call, it's because they want to communicate something to other cranes, and I know that they "know" that other creatures can hear their calls as well. There is no other way to interpret why they often "address" humans with a warning call when they come too close, and certainly why a male crane, wanting to hide the location of the nest from me, sounded a "solo duet" in the fake nesting site.

The cranes don't have to answer me when I call with my megaphone and simulate a cheeky pair of territorial cranes. I never did this in the Brook, but outside the Brook, as well as in my other project areas in Germany, Asia and North America, it was a method by which I could get many responses in a short time. However, some cranes did not respond—I respected that and did not get annoyed. But each time I managed to place my megaphone call well enough in terms of timing, volume, and direction so that a pair responded, I became joyfully excited. When they shout their duet, they communicate to all other listeners, including me, "Here we are, this is our territory, and we two are a pair."

The vocalizations of the cranes are messages to the environment as a whole, thus, for example, to me. Many people have also heard these vocalizations, but have not understood them. I have even gone a few steps further and communicated with cranes. When I receive a response to my megaphone call, this is a small, apparently simple, but basically sensational communication. A human and a crane pair are talking about two central questions of crane life: Who owns this territory, and are they mated?

The communication becomes more complex when I don't get an immediate answer, but instead either the pair flies up and comes closer, for example, or one of the partners, usually the male, flies or walks alone in my direction to look for the troublemakers. When this happens, they often emit gurr sounds that show curiosity: "What's going on, who's there? I don't know you!" I too have such sounds in my (digital) repertoire,[1] and have sometimes engaged in outright conversation, just for fun, by cooing back through the megaphone.

I'm not even sure what exactly I "said" then, and I certainly didn't understand everything the crane said, by a long shot, but it was wonderful. In the end, this type of artificial vocal communication was even used successfully in raising Whooping Cranes in Wisconsin and in flight training for reintroduction. I feel this is a validation of my work. So much more could and should be done to better understand crane communication.

I am fortunate to have understood a tiny fraction of the messages, and thus to have solved some of the riddles of the cranes, after thousands of years in which the cranes have allowed us only myths and superstitions instead of enlightenment. I am very impressed by how intelligent the cranes have proven to be—they are much smarter, more foresighted and strategic, much more self-aware and more sensitive, more eager to discover, more flexible and more communicative than was suspected for many years before, including by me.

My admiring wonder at nature and evolution grows the more animal behaviors prove to be not innate but learned, even culturally passed down through generations. The more accomplishments that can be found in animals that would once have been considered exclusively "human," the more humility, modesty and respect for nature we should develop. For it shows us that we have hitherto regarded ourselves, in our humanity, as being far too unique. The more we learn, the more we notice how little we actually know.

In my opinion, these are all good reasons to want to learn more about animals and their abilities. Above all, however, they should be good reasons to

---

[1] e. g., http://bit.ly/32QcYFG

cease any further ruthless destruction of their livelihoods (and thus our own). All too often, conservationists and biologists have a similar tendency to view animals as objects, namely, "objects of observation". They are too often merely counted, recorded and mapped. But the knowledge gathered so far about the living creatures of this earth does not describe the full dimension of their lives, but only the surface. Just like all other highly evolved vertebrates, cranes are independent living beings with a personality, with the same right to life as we humans claim for ourselves. They are just not as powerful or as ruthless as we are.

Those animal species that we regard as competitors in our drive for sole dominance are being or have been persecuted and hunted to extinction—the bear (which was exterminated in Germany), the wolf (for which shooting quotas are again now being loudly demanded in Germany) and the white-tailed eagle may serve here only as examples. Other animal species, which we do not regard as competitors, are considered either as pure farm animals, which we catch or domesticate, or as "pests", which we fight. All other animals, such as lapwings, sparrows, pine martens, hares, field hamsters, partridges or cranes, are ignored. Wetlands are drained; meadows are driven over with rollers weighing tons in spring, when the meadow birds have laid their eggs, so that the hay harvest will be easier; dead-sprayed fields ensure that skylarks no longer find food; and wildflower-free agricultural areas and clean urban rhododendron gardens with monotonous lawns mean that insects, even those that we consider "useful", are drastically decreasing because they lack habitats.

The WWF Report 2018[2] shows that, currently, only about a quarter of the Earth's land area is relatively free from human influence, and that this proportion will have fallen to 10% by 2050. The loss of vertebrates (mammals, fish, birds, amphibians, reptiles) between 1970 and 2014 averaged 60%. At the same time, the number of insects has fallen dramatically, although, so far, there are only selective figures—and they are frightening.[3]

As a depressed teenager, I started discovering and collecting bird feathers in the Ruhr area, engaged in drastic and life-threatening experiences as a chemistry student in a full protective suit at the illegal cyanide dump in Bochum, read *The Limits to Growth* soon afterwards, discussed it with commitment and

---

[2] M. Grooten/R. Almond (eds): WWF 2018 Living Planet report—"Aiming Higher", Gland: WWF.

[3] cf. https://www.nytimes.com/2018/11/27/magazine/insect-apocalypse.html, see also new, more comprehensive study results, review article here (in German): https://www.zeit.de/2019/46/artensterben-insekten-klimawandel-oekologie-langzeitstudie; source references in links are here: https://www.zeit.de/wq/2019-46#sie-zaehlen-was-stirbt, especially the publication in "Nature": https://www.nature.com/articles/s41586-019-1684-3 (all links had been visited last time on Sept 24, 2021).

decided to contribute to sustainable management with my scientific research as a chemist. As a young family man, I introduced my sons to nature from the beginning. All of this led me first to crane protection and then to crane research, and an accompanying awareness gradually set in: It is not enough to focus only on waste separation, for example, or only on saving a certain amphibian species. Humankind must not limit itself to isolated measures. We must think in broad ecological terms, integrate species protection into comprehensive nature conservation, and understand nature conservation and environmental protection as necessary elements in the protection of the earth's life-support systems. We must not make the mistake of thinking that nature conservation and environmental protection are just two different expressions for the same conservation effort.

I reject the widespread view that we humans can "manage" nature with our technology and "repair" damage without consequences or unintended side effects. We don't know nearly enough about how the ecosystems work in which we live and that surround us. We don't even know how many species there are, let alone their places, their roles, and their functions in the various ecosystems. That is why we must leave nature to utilize all of the options available to it and only help it to once again unleash its self-healing powers with careful, well-considered measures. For this, our usual nature reserves (mostly managed and used by humans) are similarly insufficient. There must be many more and larger areas where nature is allowed to be left to its own devices.

It is true that the percentage of protected areas included in the "global protected network" has risen from 9 to 15% of the earth's land area. But according to a group of researchers led by Professor Watson of Queensland University in Australia, many of these protected areas exist only on paper, because the states to which they belong lack the money and/or the will to maintain them. And a third of designated protected areas are encroached upon by intensive human use, whether mining, housing development or agriculture[4] (in the Brook, it's extensive hay farming, but really the accompanying intensive drainage, and that's in a nature preserve originally set up to regenerate something like the original wetlands that the Brook had been before humans started draining them to extract peat and to set up pastures). Only 10% of land in protected areas and national parks (i.e., only 1.5% of the Earth's land area) was completely free of human use in 2017, the Australian researcher wrote—this are the northernmost lands of Canada and Russia.

---

[4] cf. http://science.sciencemag.org/content/360/6390/788; and https://www.bbc.com/news/science-environment-44155592 (both links last visited on Sept 24, 2021).

But a comprehensive rethinking must also take place for the landscapes used by humans in forestry and agriculture: Forests must be managed organically or, even better, left to themselves. Intensive agriculture and livestock farming must be converted to organic methods. People need to be involved. For decades, in my medium-sized chemical company, I have long practiced the motto, "Think globally, act locally". A good 30 years ago, I switched a large part of our electrical consumption (we were Schleswig-Holstein's second largest electricity consumer at the time) to wind power and our own combined heat and power plant. I developed energy-saving and other environmental resource-saving products and processes and researched a completely new class of materials, which I successfully brought to market, where it makes a small contribution to lowering the use of energy and raw materials. Finally, almost 15 years ago, I bought a third of the shares of a then-small and unprofitable and small biodynamically working Demeter farm and invested there. After years of strong growth, we are now operating profitably, with more than 70 employees on about 450 ha of rented land between two locations, and are probably Europe's largest such farm with the "CSA" business model ("Community Supported Agriculture"),[5] with around 1000 members and six of our own shops in 3 towns, 4 of these shops in Hamburg. We let wildflowers bloom on 5% of our arable land. There, it hums and buzzes, even more than it does on the other 95% of our arable land which also shows a dense insect life. None of the plots that we have created, each with a single crop, is larger than five hectares (which requires very precise farming operations with GPS-controlled machinery).[6]

In addition, we follow a crop rotation system that is optimal for each soil type, such that, for example, three soil-building crops are grown over the course of 6 years, and decomposing (nutrient-withdrawing) crops are grown in the following 3 years. Over the years, our soils have developed a much deeper humus layer and have a soil life one order of magnitude more active than conventionally farmed fields; moreover, we do not discharge any toxins into the groundwater—our well water is demonstrably healthy and tasty. We can drink it without hesitation and without pre-treatment. We keep exactly as many animals (cows and pigs) as we need to fertilize our land, and all of our animals are fed with the crop yields from our fields and meadows or pastures. The fact that, at our second location, where about 240 ha of our land are very well rounded up, i.e., "lying together", at least seven to eight skylark territories can be seen and heard year after year, plus around a dozen quail territories

[5] cf. https://kattendorfer-hof.de/solidarische-landwirtschaft/ (German), for English introduction see https://urgenci.net/, both links visited on Sept 24, 2021).

[6] https://kattendorfer-hof.de/kattendorfer-hof/ackerbau/

(apart from many other bird species), is an indicator of the very positive effect our farming methods are having for biodiversity.

We humans must act quickly to stop the devastation of biotopes, the waste of soil through malnutrition (far too much meat and too much wasted food), the impoverishment of the agricultural landscape and soil erosion through monocultures, overfertilization, insecticides and pesticides, and yet still create more nature conservation and wilderness areas despite rising populations with scarce land.[7] My crane conservation and research efforts are embedded in these thoughts and ideas. Not least through George Archibald's influence, crane conservation is comprehensively understood worldwide as biotope conservation, which cannot be carried out in conflict with the people who live near crane breeding, resting or wintering areas, but only with them.

It represents a great ray of hope to see that organic farming is being introduced across the board in the Indian state of Andhra Pradesh.[8] No longer are herbicides and pesticides used, and nor are artificial fertilizers. The same applies to India's smallest state, Sikkim, in the Himalayas.[9] Since 2015, agriculture there has been exclusively organic. What is particularly surprising, and rather unbelievable for many people interested in nature conservation and organic farming, is that, in North Korea, the fields and gardens are almost exclusively organically farmed. The reason for this is that, after the Soviet Union collapsed, affecting their ability to continue to support North Korea, agricultural chemicals became increasingly scarce. Out of necessity, the nation began to revisit its original farming methods. It is largely unknown that George Archibald and his International Crane Foundation played a very helpful role in bringing modern knowledge of organic farming to North Korea in those years. This improved the quality of organic farming methods. George had brought textbooks and technical articles into the country, which were translated and disseminated along with experts who helped on the ground, and more. George and his ICF were active where there were conflicts of interest between crane conservation and agriculture. They made sure that these conflicts could be resolved. Nevertheless, with the increasing isolation of this country and the massive sanctions put in place upon it in recent years,

---

[7] cf. a study published in "The Lancet" https://www.thelancet.com/pdfs/journals/lancet/PIIS0140-6736(18)33179-9.pdf, and a discussion between the well-known US biologist E. O. Wilson and the German marine biologist A. Boetius: https://epaper.zeit.de/webreader-v3/index.html#/843443/33 (readable only with digital subscription in German), both links last visited on Sept 24, 2021.

[8] cf. https://www.zeit.de/2018/27/landwirtschaft-kosten-lebensmittel-umwelt-un-bericht (readable after registration), as well as https://www.presseportal.de/pm/7840/4017822 (both articles in German, links had been visited last time on Sept 25, 2021).

[9] cf. https://schrotundkorn.de/lebenumwelt/lesen/willkommen-in-sikkim.html (article in German, visited last time on Sept 25, 2021).

agricultural production is again critically low, in spite of increased efforts to use organic agriculturing methods.[10]

We can show our respect for nature by, for example, utilizing more considerate planning in agriculture and forestry or in infrastructure development, discontinuing drainage projects and instead rewetting previous moors and wetlands, allowing rivers and creeks to again meander and flood the river banks and riparian forests, sealing less soil, or, better yet, choosing more often to simply leave other living things alone. We would be richly rewarded—with fertile soils, clean and abundant groundwater, and diverse plant and animal life. Erosion would be stopped, devastating floods would become rarer. Our livelihoods would become healthier, and so would we.

Unfortunately, we are still a long way from that, and so, in Germany, as everywhere else, the world is by no means in order for cranes and the vast majority of wild animals. The Eurasian Cranes are officially classified as "no longer endangered", but it is a fact that, until 100 years ago, there were many cranes in all federal states. Today, after considerable effort, almost 10,000 pairs breed in Germany again, but they are spread over a few federal states in the north of Germany: The majority of cranes are found in the new federal states of Mecklenburg-Western Pomerania and Brandenburg (which can be attributed to the successful nationwide crane protection in the GDR), with quite a few also residing in Schleswig-Holstein and Lower Saxony. These cranes had migrated from the former GDR because there was not enough space for the increasing number of birds. In North Rhine-Westphalia, there are only a few cranes, in Bavaria and Hesse, to name only 2 states, none at all.

How poor is our understanding of a "rich life" and "vibrant nature" if we are content with a few thousand cranes, and consider this to be a state of affairs that is not worthy of further improvement? Wouldn't it be desirable for our landscapes to become more diverse again and for more animals to find a habitat again, separated from the almost 85 million people in Germany? By this, I do not only mean the 10,000 crane pairs, or the 600 white-tailed eagle pairs, or the 1200 peregrine falcon pairs, or other rare and not-so-rare species, but also those that are not so easy for us to tolerate in our vicinity, such as wolves and bears. And those that we do not notice so much, but are missing nevertheless, like field hamsters and partridges. In Germany, we have designated only 0.6% of our land area as national parks, only a minor fraction of which is true wilderness, where nature is truly left to its own devices.[11]

---

[10] cf.   https://www.wiwo.de/unternehmen/handel/landwirtschaft-nordkoreas-bauern-sollen-auf-bio-duenger-setzen/19473324.html (article published in 2017 in German, visited last time on Sept 25, 2021).
[11] Newspaper Die Welt as of 8.10.2019, see https://www.welt.de/wissenschaft/umwelt/plus201308558/Unterwegs-in-deutschen-Nationalparks.html (article in German, visited last time on Sept 25, 2021).

I have no illusions about how far away we, the society as a whole, are from the necessary degree of awareness for nature and biodiversity protection. However, we as citizens need not wait for the government or the EU or UN or the president to finally enact appropriate legislation. After all, every individual can do something. For example, we can get involved in nature conservation, whereby I do not mean writing letters to the editor or protest letters to Congress, but rather active participation: Planting and maintaining meadows of orchards and wildflowers, helping rewet previously drained wetlands, participating in local nature conservation projects similar to the crane protection that we started 40 years ago, the success of which can be clearly seen today.[12]

For example, you can become a member of one of the numerous organic farms that, like us, follow the CSA business model. In Germany, there are more than 200 such farms, with probably more than 25,000 members[13] who finance them—by far not enough! In France, there are many more, over 2000, with probably over 320,000 members. In Italy, there are about 1000 such farms, and in Europe, as a whole, there are probably about 4000. In North America, there are already over 13,000, by far the largest number of which are in the USA (although some of these are only run online, and there appear to be quite a few fake "community-supported agricultures" among them). Farm membership there ranges from a few dozen to well over 10,000 members, much larger than the largest European CSA-based farm with more than 1000 "eaters", our Kattendorfer Hof. And believe it or not, there are already around 2000 such member-financed farms in China, not to mention that this type of farming was invented in Japan in the mid-1960s. Even South Korea is more developed in this respect than Germany.

There is an international network[14] that promotes the concept of CSA worldwide with advice and action, as well as through the exchange of experiences. You don't have to start such a farm yourself; you can achieve a lot merely by becoming a member of the solidarity-based community of an existing farm; even more so if you initiate one in conjunction with a farm. The special thing about this construction is that it creates a win-win-win situation, i.e., a triple profit: The customers (the members of the CSA) win because they receive honest, tasty, organically produced food of regional origin at prices that are comparable to those of an organic food store. The farmers win because they no longer have to sell through the extortionate price-squeezing dairies

---

[12] More on this in the appendix.

[13] An up-to-date list of existing establishments can be found here: https://www.solidarische-landwirtschaft.org/solawis-finden/liste-der-solawis-initiativen/.

[14] cf. http://urgenci.net/

and organic wholesalers, but can sell directly and receive the trade margin themselves. In times of crisis, the members finance the farm (precisely at such moments when, as in 2018, products like milk, cheese, and vegetables are in short supply due to a lack of rain), because it is an economic community based on solidarity (many farmers are also happy when their customers occasionally or regularly help out on the farm). Thirdly, nature wins because it can recover on the organically farmed land. Soils become healthy, soil erosion is stopped, groundwater becomes clean again, insects and birds will return.

As much as many of us may be tormented by how slowly progress is being made in protecting nature and how many setbacks there are, the progress that has been made here and there should nevertheless make us optimistic. If that doesn't convince you, perhaps a look at the cranes will: Those that have once again lost their eggs or their chick will start a second brood. If that fails too, they start again the next year. As has happened for 60 million years.

Appendix

Emotions That, According to My Observations, Cranes Have or Do Not Have

The following list of human emotions (which does not claim to be complete) has been developed over the years of my occupation with these topics, with the help of various sources.[1] I will now compare which of the emotions known to us are definitely or presumably found in cranes.

In my opinion, cranes clearly showed *joy*, for example, when arriving in a territory or after the first flight day of a young bird (cf. chapters "Arrival in the Brook After Returning from Wintering Grounds: Alone or in Groups?" and "In the School of Life").

I think I observed *anger* at least once, in the "Pioneers", because of a brood loss: When one partner was searching for food, the other came flying to him from the nest, and they both marched through a group of horses, obviously aroused and with an unusually large distance between them, something they would never do otherwise—to me, the behavior appeared "angry" (chapter "Problem Solving, Ballet Courtship and the Fox Alarm: How Do Cranes Communicate with Each Other?").

---

[1] Among other sources, Joseph Ledoux, *Das Netz der Gefühle*, Hanser 1998 (original english edition: "The Emotional Brain. The Mysterious Underpinnings of Emotional Life", Simon & Schuster 1998); cf. also Dorsch, Lexikon der Psychologie, there: "primäre" or "sekundäre Emotionen", or "Unterscheidung von Emotionen nach Mees 1991", in: https://de.wikipedia.org/wiki/Emotion (English: https://en.wikipedia.org/wiki/Emotion), as well as Lexikon der Neurowissenschaft: Rüdiger Vaas, Emotionen, https://www.spektrum.de/lexikon/neurowissenschaft/emotionen/3405, Akademischer Verlag Heidelberg 2000 (German).

B. Wessling, *The Call of the Cranes*, https://doi.org/10.1007/978-3-030-98283-6

Mourning: Yes, as in "Romeo" mourning "Juliet" (see chapter "Breeding Season: A Tragic Love Story").

Interest: Yes, as in the Japanese Red-Crowned Cranes gathering to look at the raccoon dog (see chapter "Research Adventure: Red-Crowned Cranes Eavesdropping at Minus 25 Degrees; Crane Research with Armed Border Guards in the Background").

Disgust: I would definitely say yes; often, I found that one of the partners would express obvious disdain when they didn't want something that the other intended, for example, when one partner wanted to take off but the other didn't want to take off.

Fear: Yes, cranes fear foxes, sea eagles and humans to a greater or lesser extent, with definite individual differences. Cranes clearly share their sense of fear with others; in other words, they communicate with each other about this feeling: The crane that recognizes the eagle and is afraid emits the warning call that encourages all of the other cranes to defend themselves for fear of attack (on young or decrepit members of the group, not necessarily for fear of attack on themselves!).

The next stage is called "self-conscious emotions", otherwise known as "secondary emotions".[2] These include:

Pride: I think I have clearly recognized this, for example, during the fantastic courtship display, when the male of the "Sly Dogs" presented himself to his partner like the "federal eagle", the Coat of Arms of Germany (cf. chapter "Problem Solving, Ballet Courtship and the Fox Alarm: How Do Cranes Communicate with Each Other?"). Possibly, one could also classify the "bragging march", a typical activity for cranes with which territorial boundaries are often marked out, under "pride".

Shame: No, not that I could discern.

Feeling of guilt: I couldn't find any evidence of this either.

Embarrassment: Same thing. I saw nothing indicating this.

Envy: Perhaps. I have never observed food envy between partners; but if we classify the repeated attempts by cranes to conquer other territories, or the attempt by the widower "Romeo" to entice the female away from the "darned seventh pair" (cf. chapters "Breeding Season: A Tragic Love Story" and "In Search of the Cranes' Language: They Call and Thus Tell Us About Their Lives"), under "envy", this could possibly apply.

Empathy or compassion: Again, I would say yes, although Marc Bekoff and Jessica Pierce attribute empathy solely to mammals.[3] I disagree, because when

---

[2] https://en.wikipedia.org/wiki/Self-conscious_emotions
[3] Bekoff/Pierce, op. Cit. (German edition), p. 132.

the crane parents, for example, patiently demonstrate to the young how to take off and fly, they empathize with their offspring's problem of learning to fly. Also, the preparation of the playgrounds for the young in the reed-grass and sedge wetlands in the Ukraine well before hatching speaks to the ability to empathize (cf. chapter "Can Cranes Think Strategically? More Amazing Observations"), as does the strategy of the "Planners" not to leave the narrow nesting site until the chick could manage the several hundred meter hike through the overgrown forest (cf. chapter "Problem Solving, Ballet Courtship and the Fox Alarm: How Do Cranes Communicate with Each Other?"). Moreover, I have very often observed that cranes can put themselves in the place of an unfamiliar observer, can imagine their location from that observer's point of view. The story, described below, of George Archibald's Sandhill crane neighbor, the flightless "Sandy", to whom the unfamiliar parent introduces a friend, undoubtedly demonstrates empathy. Likewise, I think when we observe "shared joy" (for example, "Romeo and Juliet" upon their arrival in the spring, or the "Foxes" after the first time their offspring managed successfully to fly at least a somewhat greater distance without any problems), empathy is at play: One crane notices that the other is also happy, and then they both express their joy together. At the very least, we are dealing with *mood transfer*.[4] This is an important element of consciousness, and we can observe the transmission of moods (joy, fear, worry) more frequently in cranes.

Basically, we know little to nothing about whether and how emotions are consciously felt and reflected by cranes. After all, cranes cannot speak, and if they communicate with each other in a differentiated way, we do not understand it. In at least two events, however, I think I can also recognize this ability: The grief of "Romeo", and how he communicated it to me, and the joy of the "Foxes" at the end of their young's first day of flight, and how they shared their joy and communicated it to each other—both communications, I think, require some awareness of one's emotional state.

The following abilities are also part of consciousness and form facets of it:

Imitation: Yes, young cranes do nothing else while they learn. Additionally, the Ukrainian crane pairs that did not invent the idea of creating a playground (one pair would have been the first to do so), but adopted it as a good practical idea, also demonstrate the ability to imitate (see chapter "Can Cranes Think Strategically? More Amazing Observations").

Deception: In my opinion, cranes deliberately try to deceive their environment and potential disturbers about where their nest is, cf. the "Sly Dogs" (chapter "Problem Solving, Ballet Courtship and the Fox Alarm: How Do

---

[4] Bekoff/Pierce, op. Cit. (German edition), p. 138.

Cranes Communicate with Each Other?"); then, please remember the young Red-Crowned Crane female F15 on Hokkaido, with her story of deception in which she mated with the old territory owner first, then dispelled him shortly after their weeklong affair and replaced him with her young "boyfriend"; also see the story "Red-Crowned Cranes: Deception and Tactics" further below. According to Bekoff/Pierce, the ability to deceive requires a higher cognitive level than does honesty,[5] which we can observe directly ourselves: Children learn to deceive those people around them only over time. Young children often cannot keep a secret. We all know very well ourselves that one has to "think around the corner" when being deceptive.

Disguise: In my opinion, the ability to disguise oneself goes hand in hand with the ability to deceive and is practiced by the cranes in the "solo duet" (cf. chapter "Problem Solving, Ballet Courtship and the Fox Alarm: How Do Cranes Communicate with Each Other?").

Use of symbols: Cranes also seem to do this. First, in courtship, when they throw potatoes, stones or branches behind them, or sometimes balance a stick, whatever that symbolic action may mean. Second, the "bragging march" when defending territory is clearly a symbol instead of a real act of aggression (Fig. 1).

Pointing at something: Difficult to determine, but I noticed several instances in which parents "showed" their offspring something, i.e., pointed with their beak where there was something to eat.

Role reversal (i.e., playing other roles or putting oneself in the other's place): Difficult to judge, but there are signs, such as the fact that non-sexually mature pairs build nests, so-called "play nests"; perhaps they play at "brooding". I would also recall the behavior of the "Planners", who were able to put themselves in the world of their chick and understand that they would not be able to leave the nest area until it was 2 weeks old (cf. chapter "Problem Solving, Ballet Courtship and the Fox Alarm: How Do Cranes Communicate with Each Other?"); I judge the Ukrainian playgrounds similarly (cf. chapter "Can Cranes Think Strategically? More Amazing Observations"). Consider also the decision of the male of the "Sly Dogs" to fly behind the kink because I would not be able to see him then: He could imagine that I would be able to see him if he flew in front of the kink.

---

[5] Bekoff/Pierce, op. cit. (German edition) p. 85, cf. also V. Arzt, I. Birmelin, Haben Tiere ein Bewusstsein? ("Do Animals Have a Consciousness?") Goldmann 1995 (available only in German), among others p.265. Incidentally, these authors also refer almost exclusively to research on mammals and predominantly to animals in captivity. Birds play a very minor role in the book. Mrs. Pepperberg's parrot Alex and her work with him are described, Professor Birmelin's observations of his budgies in a spacious cage are also mentioned, but otherwise, birds are extremely underrepresented, corresponding to the lack of research, especially on wild birds.

Fig. 1 A Eurasian Crane is dancing and playing with a stick (© Carsten Linde)

Changing coalitions: Even juvenile cranes frequently form changing coalitions. I also recall here the formation of gathering and traveling groups, roosting groups, and feeding communities. All of this speaks to the ability to change coalitions. A particularly interesting example of this is the following observation: I was inside the Nienwohld Moor, on the edge of the high moor core area, and wanted to record crane calls. Just as I was about to start my megaphone to encourage one or another of the six pairs there to respond, I saw an eagle fly in, of a species that is almost never seen there: a greater spotted eagle! It perched on a tree at the edge of the high moor. Immediately, several of the territorial cranes there flew in, positioned themselves some distance from the eagle in a defensive posture (necks craned back a bit, beaks pointed up ahead, ready to stab), and made hissing threatening or warning sounds. They successfully chased the eagle away (a pity, actually, I would have liked to watch it a bit longer). This event is therefore significant and should be seen as evidence of the ability to form changing coalitions, because cranes actually defend their territory against neighbouring pairs on a daily basis at that time of year. But then, in the middle of the breeding season, they suddenly formed a coalition among themselves to drive away the eagle.

Likewise, the fact that cranes defend territories everywhere in spring and summer, but migrate, rest, overwinter and forage together in large groups on suitable areas in autumn and winter, is an indication of this ability. Here, the keyword *"cooperation"* needs to be mentioned, whereby it is still to be considered whether cooperation somehow takes place automatically, routinely (as with ants) or "only" on a case-by-case basis. The last seems to me to be the case with cranes, for example, during the "group flights" with a detour via Hamburg in spring; or when guards are sent ahead to explore roosts for the group before the whole troop flies in (cf. chapter "Can Cranes Think Strategically? More Amazing Observations").

I tend to understand the rapid development of new resting places on migratory routes described in chapter "What Can We Learn About Intelligence, Migratory Behavior, the Formation of Culture, Tool Use, and Self-Awareness in Cranes?", and even more so the development of new wintering places (such as Arjuzanx in France), as "cooperation"; cooperation that can hardly be understood without allowing for relatively differentiated *communication*. The fact that relatives are tolerated as nearest neighbours with overlapping territories, i.e., the differentiation as to who is cooperated with and who is not, also belongs here, I would say.

Related to these aspects is *helping* (after all, cooperation is when the cooperating partners each contribute something to the achievement of a common goal; helping means that only one contributes something, while the other receives something): Here, I recall the call of the hungry crane that has been forced to sit on the nest for an extended period of time; the partner that has been loitering out of the house for too long recognizes the voice of its brooding partner from a distance, recognizes the call to come home, flies over, and takes over the brooding (cf. chapter "Problem Solving, Ballet Courtship and the Fox Alarm: How Do Cranes Communicate with Each Other?").

From my point of view, the experts have overlooked the *ability to hide* as a criterion for an awareness of oneself. It has already been mentioned several times in the previous paragraphs.

Ability to form relationships: Cranes form stable relationships and are together virtually all of the time, except during breeding when the partner is foraging. The indisputable fact that cranes may separate from their partner and form a new partnership while the other partner also seeks a new mate, is strong additional evidence of the ability to actively form such relationships. In my opinion, this is also linked with the next item.

Sense of justice/fairness: Cranes take turns brooding and distribute the brooding time quite fairly among themselves, as well as feeding and guiding the chick and the adolescents; it is interesting to see that, when one partner is out for "too long", the partner sitting on the nest calls him, reminding him of his duties—and this partner does indeed return after a few minutes.

Ability to plan: Territorial and breeding site selection, nest building, anticipation of the water level, access to the later leading meadow, any problem solution (which we have discussed in previous chapters), the "playground" preparation in Ukraine, selection and changing of the wintering site and the gathering and resting places, planning of the travel route and the accompany-ing travel group—I believe that the crane's ability to plan and to solve prob-lems is the most developed of all of the criteria for an awareness of themselves.

Play: In the summer and fall and during hibernation, one can often observe cranes, especially juvenile cranes, playing quite obviously—throwing objects that they find, such as branches, plastic pieces, or potatoes, into the air, danc-ing as they do so, picking them up again, and throwing them back up or behind them. Building "play nests" could equally be considered playing.

Experiences with "Sandy", the Flightless Sandhill Crane

I don't know "Sandy" and "Katie" personally—they got their names from George Archibald, the founder and then-president of the ICF. They are North American Sandhill Cranes, and are among the few cranes in this book with human names. This exception came about only because the birds were so named by George and had practically become his "pet cranes".

Sandy was 2 years old, had lost half his right wing in an accident, and had been picked up and fostered by a farmer. He spent the winter in a barn on George's property, along with another foster, whom George called "Norman".

In early summer, George built a fence for Sandy and Norman on his prop-erty around parts of a pond created by a beaver. It wasn't until the cranes became accustomed to the area and surroundings that the fence was removed. From then on, they had full freedom of movement. They liked to eat grass-hoppers on the side of a busy highway. They had discovered this nutritious spot themselves, regularly migrating there during the day and returning to their roosting area on foot in the evening.

The following winter, all the waters froze over except for a spot near the spring at Beaver Pond, where the two cranes spent the night along with a pair of trumpeter swans. Unfortunately, Norman was killed by a coyote, Sandy disappeared, and George assumed that he was dead as well.

But 2 days later, Sandy suddenly reappeared, along with a troop of water-fowl from the area. George caught him (he couldn't fly) and returned him to the barn where he had spent the previous winter. In March, Sandy was released again and immediately disappeared. Two days later, George saw him in a cornfield. He had dyed a portion of his feathers a brownish red (see chapter "What Can We Learn About Intelligence, Migratory Behavior, the Formation of Culture, Tool Use, and Self-Awareness in Cranes?" for more on this) and he had a wild female crane with him.

George was afraid Sandy might be killed by a predator, as Norman had been, so he caught him again and put him in a chicken yard. The fence had been raised to two and a half metres so that Sandy could not jump over it.

The next day, a female crane appeared at the fence (presumably the animal that George had watched Sandy with a few days earlier). Sandy danced and trumpeted and tried to climb the fence. The following day, George saw them both together at the beaver pond—Sandy had jumped over the two-and-a-half-meter fence!

All signs indicated that the two partners were in harmony. The pair frequently called in unison, danced and ate together. But suddenly, an aggressive crane pair attacked Sandy and his mate, apparently to conquer their territory. The confrontations lasted 2 weeks, after which the female had left Sandy, and Sandy disappeared as well. But the aggressors were also no longer to be seen. Perhaps they had not liked the conquered territory after all, or the male crane had actually wanted to conquer Sandy's partner.

After about a week, by which time George and his wife thought Sandy was dead, he reported back safe and sound. Shortly thereafter, a crane pair appeared with their 1-year-old fledgling. This pair did not attack Sandy, who was clearly submissive. The four spent the night together, and at dawn the following day, the parent pair copulated.

The next day, the family was gone, and Sandy was alone again. A few days later, the family returned, stayed one night, and then disappeared again. This pattern was repeated a few times until, finally, the young female crane was left alone with Sandy and her parents departed. The two then called in unison for the first time; you could tell from the young female crane's voice that she was only a little over a year old.

In summer, the two together proclaimed the surrounding area to other cranes as their territory. "Katie," as George called the young female, would fly to the territorial boundaries, sometimes with her mate, sometimes alone, asserting with the "brag march" whose territory it was.

One day in August of that year when George was working in the garden, he suddenly heard crane flight calls in the air. Soon after, three cranes landed at Sandy and Katie's house. There were two adults and one very young crane that had just fledged and was still cinnamon brown. The adult ones were not alarmed at George's presence, but the young one was clearly nervous. The adult birds seemed to know their surroundings. Were they Katie's parents?

Katie went up to the adults; Sandy was hesitant. Finally, however, the cranes came together for a few minutes. Sandy didn't seem to like this very much.

The presence of strange cranes meant stress for him: He moved about 200 m away towards the nearest pond. From there, he called out.

Katie, however, stayed with the other cranes until evening, then flew toward Sandy, made a few low circles overhead, and landed close to him. Now, the two called in unison, the other pair answered from their pond, and the duet contest went back and forth a few times. The next morning, unison calls were exchanged again, after which the family left the area and never came back.

Sandy and Katie stayed in the wetland, even through the winter (it's worth noting that the wild female crane, fully able to fly, stayed with her partner Sandy—what compelling proof that migration behaviour is not inherited!). They were fed corn by George, which they ate from a plate in the barn. One day, the corn supply ran out, leaving the plate empty. The two cranes then marched purposefully to George's house, strode up the two steps to the deck, walked along the side of the house and peered in through the glass of the kitchen door. George and his wife were quite surprised, and, of course, they gave the cranes something else to eat first.

It is astonishing how intelligently these cranes behaved, given that they were confronted with completely different challenges than all other cranes due to Sandy's handicap. They searched for food in George's house, although they had never been fed there. They had simply observed that those who brought the food to the barn stepped out of the house through the kitchen door. Evidently, they had drawn the logical conclusion that the kitchen must be the real source of the food, and found themselves there when there was nothing left in the barn.

Another interesting aspect of Sandy and Katie's story is that Katie was apparently literally "dumped" by her parents in the territory of a single male crane. In an environment where possession of a wetland is an essential prerequisite for successful reproduction—other cranes had, after all, already attempted to take Sandy's territory—the matchmaking of their daughter was perhaps not a foolish move on the part of the parents. Unfortunately, Sandy became a victim of coyotes the following year; his inability to fly ultimately cost him his life. But Sandy and Katie provided us with some important contributions to our understanding of crane intelligence.

Red-Crowned Cranes: Deception and Tactics

Yulia Momose has often observed, from her work in Japan, that Red-Crowned Cranes are able to deceive humans. The Japanese crane protectors and researchers band some (not yet fledged) young cranes every year. In doing so, they proceed differently than German or American ringers, whose preferred method is to suddenly pounce on the young crane from several sides during such actions.

The Japanese creep up patiently and carefully from all sides, while a "general" stands on a hill with a telescope and brings the "troops" into position by radio around the crane cub. The helpers have to endure a lot in the process: the humidity that seeps through even the best protective clothing, the mosquitoes and ticks, and the boredom of having to remain motionless for hours.

The "general" usually does not see the young crane at all, but only the adults. In the case of Red-Crowned Cranes, one can often observe one of the adult birds flying away to get tasty meals (mostly the delicious buff leeches) for himself or the young from more distant pools. Experienced cranes, familiar with such banding operations from previous years, apparently realize that their chicks are being stalked for the purpose of capture, and seem to have concluded that they are the ones who are leading the staff to their offspring. So, the leader of the banding operation observes a crane apparently caring for its young, but only sees an adult bird. Then, when his troop arrives at the location where this adult bird is, it suddenly flies away—and no chick is there! He has only pretended to take care of his chick while his partner was hiding with it. Such a day represents a failure for the crane protection group, but a joyful experience for the friends of cranes, and a gain in knowledge for the crane researcher.

In most cases, however, the catchers manage to sneak up on them. Having arrived near the crane family, they can even hear the soft contact sounds of the birds. Suddenly, the little one ducks down after the adults "say" something. They slowly move away, again with eager "care" for the young one, which is, in fact, no longer with them at all. If the catchers haven't watched carefully so as to see where the young one is hiding beforehand, failure is ensured. What I find interesting here is that the sounds that we classify as "purr" contact sounds, all of which sound the same to us, can mean "follow me" or "look here" in one instance, but then become "duck, wait for us" in this critical situation. They probably sound very different to cranes, conveying different meanings each time.

Often, however, the catchers are successful, despite the numerous deceptive maneuvers. After the young crane has been caught and carefully but quickly ringed, the adult birds usually fly over the group and call out. Often, however, it's not the excited alarm calls that are heard, but soothing "purr" sounds, which—Yulia believes and hopes—may now mean, "Don't worry, we'll get back to you quickly, they'll let you go in a minute." After all, such pairs have been through this procedure a few times.

By the way, the banding project produced an interesting result:[6] It resulted in a record of when the young banded birds mated and where they had looked for territories. The males settled, on average, only 9.7 km away from their home range, the females 45.3 km away. This is in essential agreement with observations of Grey Cranes and Sandhill Cranes.

Peace-Loving Cranes?

Cranes always seem very peaceful, calm, self-confident and even majestic. They do not pick on each other, and their territorial fights do not result in the flying of feathers, let alone blood.

Why, then, did the peace movement and the UN choose the dove as their symbolic animal and not the crane, even though the crane is a symbol of peace in China? Perhaps the UN has read in specialist books that cranes sometimes destroy the nests of competitors, which doves possibly do not do, and this is why they are considered even more peaceful.

Swedish photographer Sture Tranevig has finally dispelled the illusion that cranes are peaceful. He documented an instance in which a crane regularly visited the nest of a pair of mergansers breeding in the immediate vicinity during the breeding season. When the merganser chicks hatched, the crane came by again and swallowed them. So, he knew that chicks would hatch from eggs (episodic memory!). And he anticipated that the post-hatch meal would be more interesting, or maybe he doesn't know as much about eggs as crows do, such as how to drink them. He planned and waited.

At the European Crane Conference at the end of 2018 in southwestern France near Arjuzanx, Norwegian scientists reported on initial observations and assumptions that, where cranes breed, the breeding success of waders is affected. However, these are only initial, uncertain observations. They reported that a study is being launched for the purpose of reaching more precise conclusions. However, these initial observations are consistent with what Sture Tranevig observed. In Norway, only recently—again!—have cranes begun breeding. A breeding pair even tried to do so at what is probably the most northerly location in Europe, at 70°39′ north latitude—it doesn't get much further north than that, even in Norway!

In Germany, a crane pair is said to have once kidnapped another's offspring (we experienced something similar during the rehearsal with "Operation

---

[6] K. Momose: Banding study of Red-Crowned Cranes in Japan, in: Proceedings of the VIIth European Conference on Cranes, Stralsund 2010.

Migration" in Wisconsin, as reported in chapter "We Fly Off: The Hard Way to the Migration Flight School"). The German "kidnapping pair" is said to have been without offspring for years before.

So, cranes are not infinitely peaceful, but they are much more peaceful than we humans or some apes. At least they settle their intra-species conflicts almost always without bloodshed.

USA: Human Friends and Enemies

That cranes are oriented not only aurally, but also distinctly visually, is demonstrated by the ability of captive-reared cranes at the ICF and Patuxent to recognize and differentially respond to their caretakers. Scott Swengel, the head keeper at the ICF at the time, told me a lot about this.

He himself is "hated" by the cranes. They don't like him at all. He takes their eggs, he steals their chicks before they fledge, he takes their blood—all sorts of evil activities. The fact that he sometimes brings them food does not garner him any credit from them. The reactions vary from crane to crane, with either submission or aggression being expressed to varying degrees.

Now, one might assume that the cranes recognize Scott Swengel by his size and his clothing, and have their characteristic "anti-Scott reaction" when he visits them in his work clothes. But when Scott accompanies visitor tours on Sundays, for example, the cranes recognize him even when he is standing in the middle of a crowd of people and dressed completely differently than on weekdays.

Sometimes, there are interns (mostly female) for whom the cranes seem to have a lot of sympathy. They are visibly pleased when a particular female intern stops by. However, if she appears accompanied by the veterinarian, the reaction is pure fear. The female doctor is associated with unpleasant experiences: "Now I'm going to be examined again, oh dear"—a phenomenon that is not unknown to us humans either.

All cranes seem to have the ability to recognize the gender of their human counterparts. How they determine this, the ICF employees have yet to determine. Perhaps by their voice? (That's what I assumed when they told me this.) It's been observed time and time again: Male cranes find female caregivers nicer, while females tend to make friends with male caretakers. Even young female crane, which are not raised in isolation but in direct contact with humans, prefer to follow males, and young male cranes prefer to follow the females among the caregivers. This habit is even used to make an initial sex determination of chicks. Determining the sex of cranes is not that easy. Obviously, they perceive more subtleties in us than we do in them.

Surprising New Observations and Practical Advice for Readers Who Want to Record and Analyse Bird Calls Themselves

At the beginning of this book, I mentioned that, since mid-2018, I have been living in a house that is right on the border of the Brook. So, it has become normal for me and my partner to see and hear cranes several times a day. Some of the territories are only 500–1000 m from us. The cranes fly past the house just a few dozen meters away, forage just 100 or 200 m outside of our living room window, and, amazingly, sometimes give dance performances, have territorial fights and have even allowed us to watch a copulation, all of this witnessed from our armchairs or from the terrace. We can hear the sounds of their calls ringing in through one of our windows.[7] So, it didn't take long before I could distinguish the calls by their pitch. But sometimes I had doubts: Was it really only three pairs that I heard calling here, and not four?

Thus, in the spring of 2019, I reactivated my Sennheiser microphone, first using my smartphone instead of the old minidisc equipment for the recordings, until I found that I was dissatisfied with it and got myself a small but powerful digital recording device.[8] But how was I going to analyze the recordings? I was not willing to spend an additional several hundred Euros for the licenses of professional programs like "Raven Pro",[9] "Matlab" or "mathematica", each of which requires special subroutines, simply because of my newly aroused curiosity. After a lot of research and frustrating attempts to find a cheap or even free solution, I ended up with two programs: One is "Audacity".[10] With this program, I can convert recordings from my smartphone into the more suitable *.wav format ("File"/"export"/"export as WAV"), but above all, I can edit the files (remove disturbing background noises, amplify insufficiently audible calls—"Effect => Normalize") or merge parts of different recordings.

---

[7] cf. https://photos.app.goo.gl/5arAuwHzm2CLK8tG8

[8] ZOOM H2n Handy Recorder.

[9] Developed and distributed by Cornell University (USA).

[10] For other bird species, of course, you first have to find out the frequency range in which they call, i.e., where their ground frequency range is (you do not need the overtones!). To find out, you create a spectrogram, which can also be used to identify annoying interjections that you can then delete with Audacity. Now that I no longer work with SoundForge, I create sonagrams online on the following website, unfortunately, without the free definition of the frequency range: https://academo.org/demos/spectrum-analyzer/

On the other hand, I have the program "Avisoft SASLab Lite":[11] It has gained a lot of functions in recent years. With this program, I proceed as follows: I open the wav file (either the original from my ZOOM recorder or a recording edited with Audacity), mark the call or the part of a call that interests me, and generate the "Power Spectrum", the frequency spectrum with the steps "Analyze"/"One-dimensional Transformation"/"Power Spectrum logarithmic, Evaluation Window Hamming, Channel 1/Ok". This gives a spectrum from zero to 22,500 Hz, completely unusable for my purposes in this form. So, I first put a grid over the result ("Display"/"Grid"), a setting that remains for subsequent operations. Now, I click on "Display"/"Display Range" and enter "X-Axis" at Xmin = 0.6 kHz, at Xmax = 1.2 kHz in the window (this is the frequency range in which most grey cranes call);[12] then, I click on "Edit"/"Smooth" and enter "Average over 41 points"; the result is a graph with an unbelievable number of pages, which I first convert with "File"/"Print" into a pdf of the same number of pages. Then, I print page 1 of this pdf file in horizontal orientation on paper, so that the DIN-A4 format is fully filled. Now, I can put the printouts of different recordings' analyses on top of each other or hold them in front of a lamp: The "peaks and valleys", if they are calls from the same pair, must be at about the same frequencies. Deviations of about five hertz are normal, as are differences in intensity (loudness) at each frequency; after all, cranes are not machines, but living creatures. The "mountains" in the spectra show the frequencies with which a respective crane calls.[13]

As of March 2019, territorial occupations were obviously not yet complete. Some pairs were breeding, others were still courting and mating, but they had to fend off various envious intruders who also wanted into their territory. In my immediate neighborhood, there are two territory owners who apparently tolerated each other (and bred in April), but staked out boundaries, in an area that used to have only one pair living in it (first "Romeo and

---

[11] http://www.avisoft.com/downloads/

[12] For other bird species, of course, you first have to find out the frequency range in which they call, i.e., where their ground frequency range is (you do not need the overtones!). To find out, you create a spectrogram, which can also be used to identify annoying interjections that you can then delete with Audacity. Now that I no longer work with SoundForge, I create sonagrams online on the following website, unfortunately, without the free definition of the frequency range: https://academo.org/demos/spectrum-analyzer/

[13] Two examples of two different crane pairs: https://www.bernhard-wessling.com/power-spectrum-examples

Juliet", then the "Sly Dogs"). There are currently at least two other pairs also staking claims here.

Right next to it, another pair occupies the territory that is least visible and has been used since the mid-1980s; there is probably a second pair there as well.

In the centre of the brook, the conditions similarly seemed to lack clarity, because, on the different days of my recordings, I altogether registered five pairs that called duets from there. There were, at most, probably two or three pairs that occupied territories there and drove away the other pairs.

At the edges of the Brook, the situation remains unclear, which is quite normal at the beginning of a season. And for someone like me, who has now decided that he wants to get to know the cranes personally again after all, who observes their constant and changing habits, the situation is particularly confusing, because many calls from one area (from which you hear something, but mostly see nothing) from different days do not match. Instead, as I have fanned out the initial 50-plus, now 250-plus call recordings, I have found that quite different pairs were calling from one territory on different days.

While the usual observer of a territory says: "Look, the crane pair is back!", I ask: "Which pair? Who actually lives here? I've heard three different pairs unison calling from the same corner!" That will only become clear in the course of the season, at the latest, when breeding begins—if breeding occurs at all in the respective territory, because, if there is a dispute over a territory, a crane pair often does not find enough peace to build a nest, mate often enough, lay eggs and incubate. In total, I identified 29 pairs in the Duvenstedt Brook plus 4 in Hansdorf Brook by their voices during the 2019 breeding season (and that was in a single season!), but probably only ten or twelve pairs occupied a suitable territory and at least attempted to breed, which is quite a lot for this area, as it is particularly crowded in the centre of the Brook.[14] From experience, breeding success declines when too many pairs are competing in a confined space. And it does really seem to get crowded: On one day in May, I saw at least 65 cranes in a large meadow in the center of the Brook, flying in one after another in three larger groups in the late afternoon. Together with the territorial pairs (probably about fifteen at that time), there were about one hundred cranes in the Brook! This is a great success of many years of nature conservation, and fortunately not unique, because we can observe similar developments in many places in Western and Northern Europe.

In the following year (2020), the situation became even more confusing, as we had lots of visiting cranes, partially up to 100 in a group in the Brook

---

[14] https://www.bernhard-wessling.com/karten

center. But another event showed the value of bioacoustic monitoring, if you want to learn more about the cranes' behaviour:

In May 2019, I noticed in my close neighbourhood that a new pair was trying to occupy a territory. This is very late, certainly much too late for breeding. Sure, I recorded its unison call, but did not take any special notice, because during summer, it was not there (not to be heard, not to be seen). Then, in late October, it suddenly arrived again, announcing itself by unison and guard calls. It was subsequently gone again during winter. The following year (2020), this pair, which I could recognize from their tooting voices (and from its calls' power spectrum), occupied a territory in my neighbourhood from the beginning of the season. They did not breed. They were still much too young. The male's tooting told me that his voice was at least cracking, but the female's voice was also not yet clear. Their voices matured during 2020 and were clean and clear in 2021. The reason why they were nevertheless easily able to conquer a territory was because two formerly established pairs, present in 2019, had not been there. This shows how actively and early young cranes are looking for a territory, far before they are sexually mature. They tried to breed in 2021, but failed.

Those readers who are considering whether they might want to pursue such or similar studies as amateur bioacousticians themselves (rather than just occasionally playfully listening in, which can itself already be very rewarding) should realize up front: If you really want to find out something about the individuals whose calls you are recording (regardless of what species of bird it is), you will have to invest some time. It's not enough to record a handful of calls. You need (depending on the number of individuals you want to identify) a few dozen to a few hundred calls from several days of the season, because birds are very dynamic. They interact with their neighbours and their environment, and nothing is the same as it was 2 days earlier.

So, you have to deal with the adversities of recording technology, weather, the all-too-early time of day, and the vagaries of the birds themselves (who rarely call when you're ready, but especially like to call when you're fiddling with the equipment). You have to deal with the intricacies of the programs, especially the printer. Additionally, when you finally assemble all of the printouts that you want to compare, the apparently simple question turns out to be a rather confused puzzle: Who actually flew from where to where and called at completely different places? Or: Wasn't it already clear in this one territory who the territory owners are, why is a completely different pair calling here now without experiencing any pushback?

The reward for all of this effort, and the compensation for the frustrations of having solved the puzzle, is, first of all, a better understanding of what one has been able to observe or hear. Included in the reward are the new questions that open up when you have deciphered a puzzle, as they did for me in March 2019: There is a pair whose call raises doubts in my mind on repeated listening—after all, no duet sounds like that! The spectrogram[15] looks like two males calling in duet (the shape of the duet parts of males and females are radically different); could this be? So far, I had only heard reports of same-sex relationships among cranes from stories about those kept in zoological gardens. But it could also be that there are two males of adjacent pairs displaying their territory in a solo "duet", while the females are in another part of the territory, and therefore are not calling. Originally, no one believed me that mated male cranes occasionally call their part of the duet alone, until I was able to prove it with my recordings a few years later. That year, I had already recorded a crane calling alone several times. But two at the same time, standing side by side? Very strange, quite a new observation, even if they weren't a gay couple.

Then, there is the case of the two neighbouring crane pairs (only audible, but not visible) in which the females have almost identical spectra,[16] but the males are clearly different. The two calls for which I noticed the acoustic similarity of the females came from very close proximity, only about 15 minutes apart: Are the females closely related (twins?), or is one the daughter of the other? Or did I hear one of the fastest new matings ever, with the female subsequently calling first with one, then with the other male in practically the same place in a duet? All three answers would be surprising, because, so far, I myself and others have only observed male offspring of crane parents breeding in the immediate vicinity of their home range (I have inferred this from the similarity of the male part of the spectra, while others have done so from ringing projects). But even more surprising would be a mate change within a period of 15 minutes and in the same location. Since the male from the second unison call called with another female in subsequent weeks, and the female from both calls called with the first male on the later recording days, "speed-dating" is the most likely explanation for me.

Another reward for the effort of recording bird calls is that bioacoustic observation, as opposed to visual, more accurately depicts dynamics: I mentioned earlier that 29 crane pairs introduced themselves to me through their calls in the spring of 2019, and also the most recent series of events with the young pair in my neighbourhood in 2020/21. But the reward of all of the

---

[15] cf. https://www.bernhard-wessling.com/two-males-unison-call spectrogram generated online at https://academo.org/demos/spectrum-analyzer/; for a better view, you need to generate a *.jpg and zoom into it.

[16] cf. https://www.bernhard-wessling.com/speeddating-2019-03-06

effort and time spent goes deeper than that: You develop a growing closeness to the birds that you observe and record. For me, at least, it is like this: The more often I hear my immediate neighbours and the other cranes in the Brook, record their calls, analyse them, print them out, and compare them, the more familiar they become to me, almost like friends. I consider it to be luck, a great gift, that, in this way, I can hear and decode some of the vocal messages of the cranes—and the calls are certainly nothing less than messages! I am allowed to hear and decode them, to get to know some of the cranes personally and to recognize them.

My experiences with the cranes always make me aware that I myself, like all of us, am but a teeny-tiny part of the cosmos. In my chemical research, I went very deep into some areas and, at the same time, knew how much I didn't know, how much I didn't understand (as is the case with all of the other people worldwide who do research in these areas). As a chemist with a general interest in the natural sciences, someone who likes to delve into physics, astronomy, genetics or ecology, it is always clear to me that we humans only know and understand a tiny section of our world—and perhaps much of what we think we know today will be considered outdated in a few years or decades later.

When I listen to the cranes once again, I become very close to them, experience myself as part of nature and gain awareness: The world around us is full of big and small secrets. When we succeed in deciphering some of them, they tell us: We humans have every reason to be modest and to behave respectfully towards nature.

Printed in the United States
by Baker & Taylor Publisher Services